RAND

Assessing Gas and Oil Resources in the Intermountain West

Review of Methods and Framework For a New Approach

Tom LaTourrette, Mark Bernstein, Paul Holtberg, Christopher Pernin, Ben Vollaard, Mark Hanson, Kathryn Anderson, Debra Knopman

Prepared for
The William and Flora Hewlett Foundation

RAND Science and Technology

The research described in this report was conducted by RAND Science and Technology for The William and Flora Hewlett Foundation.

Library of Congress Cataloging-in-Publication Data

Assessing gas and oil resources in the intermountain West : review of methods and framework
for a new approach / Tom LaTourrette ... [et al.].
    p. cm.
    "MR-1553-WFHF"
    Includes bibliographical references.
    ISBN 0-8330-3178-3
    1. Natural gas—Great Basin. 2. Petroleum—Great Basin. I. LaTourrette, Tom, 1963–

TN881.A1 A74 2002
553.2'8'0979—dc21

                       2002069686

RAND is a nonprofit institution that helps improve policy and decisionmaking through research and analysis. RAND® is a registered trademark. RAND's publications do not necessarily reflect the opinions or policies of its research sponsors.

© Copyright 2002 RAND

All rights reserved. No part of this book may be reproduced in any form by any electronic or mechanical means (including photocopying, recording, or information storage and retrieval) without permission in writing from RAND.

Published 2002 by RAND
1700 Main Street, P.O. Box 2138, Santa Monica, CA 90407-2138
1200 South Hayes Street, Arlington, VA 22202-5050
201 North Craig Street, Suite 102, Pittsburgh, PA 15213
RAND URL: http://www.rand.org/
To order RAND documents or to obtain additional information, contact Distribution
Services: Telephone: (310) 451-7002; Fax: (310) 451-6915; Email: order@rand.org

Preface

This report, along with an abridged version released as an Issue Paper
(LaTourrette et al., 2002), is an interim report from a project addressing gas and
oil resource assessments in the Intermountain West. The objective of this work is
to propose, develop, and apply a methodology for assessment that includes
additional economic and environmental considerations. This interim report
describes a set of criteria that can be applied to technically recoverable gas and
oil resource assessments that would allow policymakers to better understand the
economic and environmental implications of federal land use decisions. Because
of the inherent uncertainty in making hydrocarbon resource assessments,
building a comprehensive methodology that includes economic and
environmental considerations is challenging. In the next phase of the project we
plan to more fully develop this assessment methodology and then apply it to
Intermountain West basins.

Given the challenge of developing such a methodology, as well as its relevance to
the current debate on energy policy, we believe that it was important to release
this interim report. By doing so, we have created the opportunity to gather
additional feedback on our proposed methodology as we proceed with the next
phase of work. We welcome comments on this report from interested readers;
please direct comments to Tom LaTourrette at rockies@rand.org.

RAND Science and Technology

RAND is a nonprofit institution that helps improve policy and decisionmaking
through research and analysis. RAND Science and Technology (S&T), one of
RAND's research units, assists government and corporate decisionmakers in
developing options to address challenges created by scientific innovation, rapid
technological change, and world events. RAND S&T's research agenda is
diverse. Its main areas of concentration are science and technology aspects of
energy supply and use; environmental studies; transportation planning; space
and aerospace issues; information infrastructure; biotechnology; and the federal
R&D portfolio.

Inquiries regarding RAND Science and Technology may be directed to:

Steve Rattien
Director, RAND Science and Technology
RAND
1200 South Hayes Street
Arlington, VA 22202-5050
703-413-1100 x5219
www.rand.org/scitech

Contents

Preface iii

Figures vii

Tables ix

Summary xi

Acknowledgments xxi

Abbreviations xxiii

1. INTRODUCTION 1
   Background and Objectives 1
   Approach 3

2. TECHNICALLY RECOVERABLE GAS AND OIL RESOURCES 6
   Different Definitions of Terms 7
   Assessment Methodologies 9
      Reserve Appreciation 9
      Undiscovered Conventional Resource 10
      Nonconventional Resources 11
   Results for the Lower 48 States 12
   Results for the Rocky Mountain Region 15
   Resource Assessment Evolution 19

3. LEGAL ACCESS TO RESOURCES IN THE
   INTERMOUNTAIN WEST 22
   Methodology 23
   Results 26
   Recommendations 31

4. ECONOMICALLY RECOVERABLE RESOURCES 33
   Methodology 33
   Results 35
   Recommendations 38

5. INFRASTRUCTURE 41
   Overview of Infrastructure Components 41
      Water Disposal 42
      Compression 42
      Gathering System 43
      Processing 44
      Transmission Pipelines 44
      Roads 45
   Including Infrastructure in Resource Assessments 45
      Infrastructure Requirements 45
      Other Considerations 47
      The Viable Resource 48

vi

6.   ENVIRONMENTAL CONSIDERATIONS   49
      Overview of Environmental Impacts   50
          Exploration and Development   55
          Production   56
          Maintenance   57
          Accidents   58
          Waste Disposal   59
          Orphan Wells   59
          Transport   60
      Environmental Considerations and the Viable Resource   60

7.   CONCLUSIONS.   63
      Implications of Viable Resource Approach   63
      Potential Results.   64
      Future Work   65

Appendix
      TECHNICALLY RECOVERABLE RESOURCE
      ASSESSMENT SPECIFICATIONS   67

References   73

Figures

S.1. How Viability Criteria Affect the Available Resource xiii
S.2. Potential Effect of Viability Criteria on Gas Resources xviii
1.1. Effect of Viability Criteria on the Available Resource 3
2.1. Hierarchy of Categories to Describe Resources 7
2.2. Comparison of Lower-48 Technically Recoverable
      Natural Gas Resource Assessments 14
2.3. Rocky Mountain Resource Regions 15
2.4. Comparison of Rocky Mountain Region Technically Recoverable
      Natural Gas Resource Assessments 16
2.5. Basins Containing Coalbed Methane Deposits 18
2.6. Basins Containing Tight Sandstone Gas Deposits 18
2.7. Historical Gas Resource Estimates 20
3.1. Effect of Including Proved Reserves on Access Levels in the
      Rocky Mountain Region 29
3.2. Effect of Including Non-Federal Lands and All Gas Resources on
      Access Levels in the Greater Green River Basin 30
4.1. Economic Recoverability as a Function of Cumulative
      Production. 34
5.1. Framework for Estimating Infrastructure Requirements 46
7.1. Potential Effect of Viability Criteria on Gas Resources 65

Tables

2.1.   Comparison of Lower-48 Technically Recoverable Resource
        Assessments   13
2.2.   Comparison of Rocky Mountain Region Technically Recoverable
        Resource Assessments   17
3.1.   Classification of Lease Stipulations and Effect on Gas Drilling   24
3.2.   Reported Natural Gas Access Restrictions in the
        Rocky Mountain Region   27
3.3.   Reported Natural Gas Access Restrictions in the
        Greater Green River Basin   27
3.4 .   Natural Gas Drilling Opportunities in the
        Greater Green River Basin   28
4.1.   Economically Recoverable Oil and Gas in the United States
        (USGS)   37
5.1.   Infrastructure Components, Cost Items, and Issues Specific to the
        Rocky Mountains   42
6.1.   Potential Environmental Impacts from Oil and Gas Extraction   51
A.1.   Comparison of Resource Assessment Specifications   67
A.2.   Comparison of Resource Categories   70

Summary

The availability of gas and oil resources in the Intermountain Western United States has become the subject of increased debate in recent years. Several studies have concluded that substantial amounts of gas and oil resources in the region are inaccessible because of legally restricted access to federal lands (e.g., National Petroleum Council, 1999; Advanced Resources International, 2001). Some stakeholders have reacted to the studies by calling for reduced access restrictions, while others have called the studies flawed and support continued restrictions. The debate has sparked renewed interest in the process of assessing hydrocarbon fuel resources.

This report is part of an energy initiative by the Hewlett Foundation. In this effort the foundation asked RAND to:

- review existing resource assessment methodologies and results
- evaluate recent studies of federal land access restrictions in the Intermountain West
- consider a set of criteria that can be used to define the "viable" hydrocarbon resource, with particular attention to issues relevant to the Intermountain West
- develop a more comprehensive assessment methodology for the viable resource
- employ this methodology to assess the viable resource in Intermountain West basins.

This is an interim report that focuses on the first three tasks.

Key Policy Questions Require More Information Than Provided by Traditional Assessments

The goal of traditional resource assessments is to estimate the nation's potential supply of natural gas and oil resources. In this report, we examine four recent assessments (U.S. Geological Survey National Oil and Gas Resource Assessment Team, 1995; Minerals Management Service, 2000; National Petroleum Council, 1999; Potential Gas Committee, 2001). Although the assessments vary, they each indicate that the Intermountain West contains substantial natural gas and oil resources. Traditional resource assessments, however, are intended to estimate the "technically recoverable"[1] resource, which does not reflect the amount of resource that can realistically be produced. Technically recoverable resource assessments, by design, make no assumptions about whether or not the resource will be developed, and resources are evaluated regardless of political, economic, and other considerations. The distinction between the technically recoverable resource and that which is likely to be actually produced is important when confronting questions about the potential benefits and impacts of increased natural gas and oil exploration and production.

The amount of resource that is likely to be produced depends on a number of considerations. The criterion that a resource be technically recoverable is only one of several that are relevant to determining if that resource is, in fact, recoverable. Legal access restrictions, as it turns out, may not always be the pivotal factor for actual resource development, because other factors may play greater roles in determining if a resource is recoverable. Three key factors are:

- exploration and production costs (those incurred in getting the resource to the wellhead)

- infrastructure and transportation costs (those incurred in getting the resource to the market)

- environmental impacts.

The wellhead and infrastructure costs are relevant, because, when compared to the revenue expected from the resource being considered for development, they determine whether it is economically feasible to proceed. Environmental impact can be treated in a similar manner by characterizing different levels of impact and allowing policymakers to consider effects at different levels. For policy purposes, these three factors could add significant value to resource assessments. The resource that satisfies this more expansive set of criteria has a reasonable likelihood of actually being developed and produced. We call such a resource the "viable" resource.

The cumulative effect of these additional factors on the available resource is shown conceptually in Figure S.1. The application of each additional criterion—wellhead economics, infrastructure economics, and environmental

---

[1]The technically recoverable resource refers to the amount that is estimated to be recoverable given certain assumptions about technical capabilities. In practice, the definition of the term "technically recoverable" is unclear and is inconsistently applied among the different assessments. A large part of the differences between existing resource assessments results from differing assumptions as to what constitutes a technically recoverable resource.

RAND*MR1553-S.*

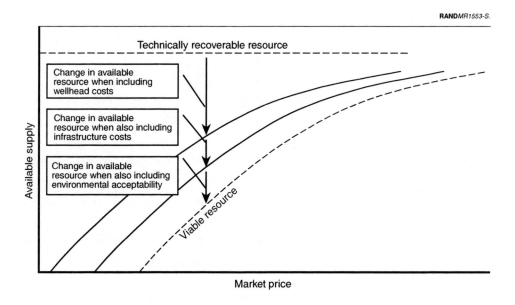

Figure S.1—How Viability Criteria Affect the Available Resource

acceptability—successively reduces the amount of a resource that might be available at a given market price. Note that the curve for environmental acceptability is conceptual only—we do not propose to calculate environmental costs. Rather, we intend to estimate the amount of economically recoverable resource that can be extracted within a given level of environmental impact.

These three factors reflect well-known and often cited issues that determine the availability of gas and oil resources. Aspects of these issues have been addressed to varying degrees in previous studies (e.g., Vidas et al., 1993; Attanasi et al., 1998; National Petroleum Council, 1999). However, they are generally not all considered in resource assessment methodologies.

Limitations of Existing Access Restriction Studies

Existing approaches to understanding resource availability have focused on legal access restrictions on federal lands. Ongoing efforts have been spurred largely by the Energy Policy and Conservation Act of 2000, which directs federal land management agencies to assess the energy potential of public lands and identify impediments to its development. As a result, considerable effort is being expended on quantifying the amount of gas and oil resources underlying federal lands that is subject to various forms of access restrictions.

The recent debates over access to natural gas in the Intermountain West have centered largely on the conclusions made in two studies. The first, conducted as part of the latest National Petroleum Council natural gas study (National Petroleum Council, 1999), addresses the entire Rocky Mountain Region; the second, prepared for the U.S. Department of Energy (Advanced Resources International, 2001), focuses on the Greater Green River Basin in southwestern Wyoming and northwestern Colorado. In their effort to identify impediments to energy development, these studies make some important assumptions that have implications for the impact of access restrictions on the available gas resource. These assumptions deal with economics, the resource base considered, restriction enforcement, technology, infrastructure, and drilling schedules. As calculation of access restrictions continues to be a component of policy guidance (studies of additional basins are under way), these assumptions should be closely examined and modified where necessary to provide an unbiased and consistent view of the impact of access restrictions in the broader context of economic constraints and the non-federal resource base.

Building Comprehensive Resource Assessments

For making informed decisions, policymakers need to know how much resource is available, at what cost, and with what impact. Therefore, rather than focus on the amount of resource that is unavailable as a result of land access restrictions, we propose an approach of determining the viable resource: that which is available when considering wellhead costs, infrastructure costs, and acceptable environmental impact.

Wellhead Costs

Wellhead costs vary depending on a deposit's geologic characteristics, depth, and production characteristics. Estimating economic recoverability involves balancing these costs with anticipated resource revenues to determine if it would be economically logical to proceed with production (e.g., Vidas et al., 1993; Attanasi, 1998). The standard costs that need to be included when considering economic recoverability are:

- exploration and development drilling
- well completion
- lease equipment
- operations and maintenance
- taxes and royalties
- return on investment.

Incorporating these costs can reduce the amount of gas and oil resources that is economically viable for production in the foreseeable future. There remains considerable uncertainty about the economics of gas and oil recovery in the Rocky Mountain Region and studies are ongoing. However, based on the U.S. Geological Survey results, adding the economic criterion alone would rule out, in the near term, the recovery of a large fraction of the gas resource in the Green River Basin that would otherwise be deemed technically recoverable (Attanasi, 1998). It is important to note that technological improvements and changing economic conditions will alter these estimates over time.

In updating evaluations of the economically recoverable resource in the Intermountain West, improvements can be made to the standard economic models to help tailor our economic evaluation to account for some of the characteristics of the region and to improve the accuracy of economic modeling of resource development.

The first is to use data that reflect the region of interest. Costs of gas and oil development in the Rockies can vary considerably depending on the location and characteristics of each basin. However, cost data are generally presented either by state or by a larger region, a practice that impairs the accuracy of the cost estimates.

The second is to account for the high abundance of nonconventional gas in the Rockies. One of the primary distinctions of the Rocky Mountain Region is the very high fraction of undiscovered gas that is contained in nonconventional formations.[2] This distinction is expected to impact costs for well completion, lease equipment, and operating costs. While existing efforts attempt to account for these higher costs by including nominal correction factors, the aggregate cost estimates may still underestimate the real costs of developing Rocky Mountain gas and oil.

The high fraction of nonconventional deposits may also influence drilling success rates. The drilling success rate is the fraction of drilled wells that are productive and influences the total number of wells that must be drilled. The rates used in existing evaluations reflect regional averages of existing wells and are thus biased toward conventional deposits.

Other unique aspects of the Rocky Mountain Region that may further influence the costs of resource extraction include the steep and rugged terrain, remote locations, low-quality gas, and shallow formations.

---

[2]Nonconventional resources in the Rockies include low-permeability (tight) sandstone and coalbed methane.

Infrastructure Costs

Much of the gas and oil resource in the Intermountain West cannot be developed without constructing additional pipeline, processing, and road infrastructure. While resources may still be economically recoverable when these additional costs are accounted for, in some cases the infrastructure requirements may prevent an otherwise attractive development from proceeding. The availability of infrastructure thus represents an important criterion for defining a resource as viable.

Typically, resource assessments do not consider infrastructure requirements. Capital expenditures and operating costs for infrastructure are thought to be comparatively high in the Rocky Mountain Region, given a lack of infrastructure relative to other regions. If new infrastructure is required, the additional costs could be more than 50 percent of the wellhead costs.

Primary infrastructure components include gathering systems, which connect wells to gas processing plants; gas processing plants, the number of which depends on the size and type of deposit; and long-haul transmission lines. The infrastructure requirements and costs depend on a number of factors, including the number and distribution of wells, well pressures, flow rates, and recovery rates, resource characteristics, and type of geological formations, all of which can be highly variable.

Several complicating factors in the Rocky Mountain Region may increase infrastructure requirements and costs. These factors include the remoteness of existing pipeline infrastructure, particularly transmission pipeline; the rough terrain, unstable soil, and icing in colder climates; the extensive water disposal requirements associated with coalbed methane deposits; and the need for extensive compressor capability to transport the low-pressure gas from nonconventional deposits.

Infrastructure costs can be assessed for different locations and ultimately parameterized in terms of a few key variables. Based on these variables, the costs can be scaled for varying distances from transmission pipelines. Beyond specific distances, development will no longer be viable.

Environmental Impact

Finally, it is important to evaluate the potential environmental impacts of exploration and production. Our proposed approach is to classify lands according to their existing environmental conditions. Individual indicators could track a spectrum of impacts, including air quality, water quality, soil conditions, hazardous materials, protected species, migration patterns, vegetation habitats, and land use changes. These conditions can be categorized and mapped to help policymakers (a) understand the spatial distribution of sensitive environmental areas within a total resource area and (b) given some acceptable level of environmental impact, select which areas are best suited to development.

Oil and gas extraction activities are regulated to mitigate environmental impacts associated with air, water, solid waste, and hazardous waste. Regulation, however, does not necessarily prohibit projects with significant environmental impacts. The potential environmental impacts of oil and gas extraction begin with the construction of the drilling apparatus, service roads, and pipelines. Solid waste, hazardous waste, and large volumes of wastewater are then generated during construction, operation, and abandonment of the project, with potential implications for regional air and water quality. There are also the rare but potentially serious effects of accidental spills and blowouts. Such disruptions could adversely affect complex ecosystems.

Potential environmental impacts may also extend beyond ecological resources to include impacts on historical, anthropological, paleontological, and societal resources. An additional potential impact with great public interest in scenic areas such as the Rocky Mountains is the aesthetic impact on landscapes. Introduction of machinery, development of roads, and the denuding of vegetated landscapes to support extraction often carry aesthetic implications.

Comprehensive Assessments Will Add Value to Policymaking

There are two primary motivations for conducting more-comprehensive assessments of the viable resource. First, it will refocus and broaden the current debate over access to federal lands. There continues to be much debate about the amount of gas and oil resources in the Intermountain West that is subject to various access restrictions. This debate focuses on the technically recoverable resource and addresses only federal lands. Policymakers and the public would benefit from a more comprehensive understanding of the broader implications of economic and environmental constraints on the availability of federal and non-federal resources. A debate about access restrictions alone does not illuminate the discussion.

Second, it would be prudent to have a better understanding of the economic costs, infrastructure requirements, and environmental impacts of increased production as policymakers consider changes in energy policy and land management practices.

At present, it is possible to make only a first-order estimate of the effect of some of these viability criteria on the amount of gas that could be viable in the Rocky Mountain Region. Estimates of economic recoverability in the Rocky Mountain Region are uncertain and studies are ongoing. Figure S.2, based on data from the U.S. Geological Survey economic analysis (Attanasi, 1998), indicates that the economic recoverability criterion alone can substantially reduce the amount of gas that is viable for extraction. At a wellhead price of $3.34 per thousand cubic feet of gas (equivalent to $30 per barrel of oil), less than 20 percent of the technically recoverable gas in the total Rocky Mountain Region is economically recoverable, and only 5 percent of the technically recoverable gas in the Greater Green River Basin is economically recoverable. Note that these results do not reflect RAND's analysis. In particular, the U.S. Geological Survey results reflect technology current as of the early to mid-1990s and do not account for expected future technology improvements. The costs of exploring and developing gas and oil deposits in the Rocky Mountain Region are decreasing with technological advances. Our economic analysis will use different data and assumptions and may produce different results.

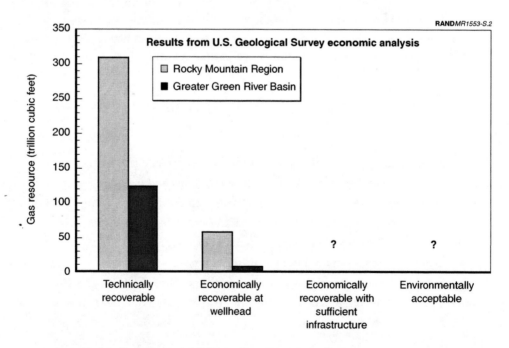

Figure S.2—Potential Effect of Viability Criteria on Gas Resources

This report lays the foundation for determining the viable gas and oil resource. The next step will be to apply this methodology to estimate the viable resource in individual basins. RAND will begin this effort by analyzing the Green River Basin. The analysis will specify the relationships among gas and oil deposits, technological options, economic costs, infrastructure requirements, environmental impacts, and other variables to allow for a comprehensive assessment of the viable gas and oil resource.

Outputs will be presented both numerically and spatially (in the form of Geographic Information System maps that show the amount and location of resources that satisfy the various viability criteria). Such an output will provide a useful way to characterize the viable resource in the context of many important variables, such as deposit types, well locations, existing and needed infrastructure, environmental sensitivities, topography, and other relevant spatial attributes. This method of conducting and presenting resource assessments would be a significant enhancement over present practice.

Acknowledgments

The authors gratefully acknowledge Robert Hugman and E. Harry Vidas (Energy and Environmental Analysis, Inc.), Peter Morton (The Wilderness Society and University of Denver), and Donald Snyder (RAND) for formal reviews that substantially improved this report. The report also benefited from discussions with Emil Attanasi (U.S. Geological Survey), John Eagleton (El Paso Field Services), Alan Wiggins (Conoco), and Blaise Pool and Paul Trousil (El Paso Western Pipeline Group). We also thank John Godges for assistance in organizing the report and Lisa Sheldone for technical support.

Abbreviations

ARI	Advanced Resources International
BB	billion barrels
bbl	barrel
BLM	Bureau of Land Management
Btu	British thermal unit
CSU	Controlled Surface Usage
EIA	Energy Information Administration
EPA	Environmental Protection Agency
EUR	Estimated Ultimate Recovery
GIS	Geographic Information System
mcf	thousand cubic feet
MMS	Minerals Management Service
NGL	natural gas liquid
NPC	National Petroleum Council
PGC	Potential Gas Committee
psi	pounds per square inch
SLT	Standard Lease Terms
tcf	trillion cubic feet
TL	Timing Limitations
USGS	United States Geological Survey

1. Introduction

Background and Objectives

The amount and availability of gas and oil resources in the Intermountain Western United States have become the subjects of increased interest in recent years. Recent national resource assessments indicate that the Intermountain West may be relatively rich in hydrocarbon resources, particularly natural gas (National Petroleum Council, 1999; U.S. Geological Survey National Oil and Gas Resource Assessment Team, 1995; Potential Gas Committee, 2001). Roughly two-thirds of this resource is located under federal lands, some of which is subject to access restrictions. This has motivated a number of recent studies addressing access to gas and oil resources in the Intermountain West (Advanced Resources International, 2000, 2001; National Petroleum Council, 1999; Barlow and Haun, 1994; DuVall, 1997). These studies conclude that substantial amounts of the resources are inaccessible for production because of various legal restrictions on access to federal lands.

These studies have elicited divergent responses from different sectors. Industry has called for reduced restrictions. Stakeholders representing environmental and recreation interests have called the studies flawed and incomplete (Morton, 2001; Defenders of Wildlife, 2001). In particular, these stakeholders argue that the impact of the legal access restrictions on the potential resource base is much smaller than the studies claim, because much of the legally restricted resources could never be developed anyway, as they are already inaccessible for other reasons.

This debate has sparked renewed interest in the process of assessing hydrocarbon fuel resources. In particular, the debate raises questions about the ways that resources are defined and about the methods used to conduct resource assessments. A large part of the difference of opinion regarding the impact of legal access restrictions stems from the fact that the terms of the debate are incompletely defined, poorly understood, and inconsistently used. A fundamental problem permeating discussions of access to resources is that the existing resource assessment methodologies appear not to be broad enough for the purpose of making informed policy decisions regarding how much resources are actually available at what cost and with what impact.

In general, the resource assessments entail gathering geologic, geophysical, engineering, and other types of physical evidence in conjunction with using various statistical methods to estimate the amount of crude oil and natural gas that may become available in the future. Resource assessments do not represent inventories of known quantities, but rather attempt to estimate the amount of as yet uninvestigated and undiscovered resources that may be added to inventories of proved reserves.

A critical aspect of resource assessments is the decision as to which resources to count. Because the assessments focus on undiscovered resources, confidence in their existence decreases and uncertainty grows as ever-more speculative resources are considered. Thus, unbounded inclusion of the most speculative resources serves little purpose for informing policy.

Most assessments attempt to report the "technically recoverable" resource, meaning that amount estimated to be recoverable given some assumptions about technical capabilities. In practice, the exact meaning and application of this criterion is unclear. For example, the U.S. Geological Survey assessment includes "resources in accumulations producible using current recovery technology," while the National Petroleum Council considers resources that will *become* recoverable in the future given an assumed rate of technological advancement. In each assessment, the application of this criterion is made through the judgment of an analyst or committee.

Uncertainty aside, the criterion that a resource be technically recoverable is only one of several that are relevant to understanding the available resource base. From a policymaking perspective, several other factors—such as the costs involved, the infrastructure requirements, and environmental impacts—play equally important roles in determining how resources should be assessed. These three factors reflect well-known and often cited issues that influence the availability of gas and oil resources. Aspects of these issues have been addressed to varying degrees (e.g., Vidas et al., 1993; Attanasi et al., 1998; National Petroleum Council, 1999). However, they are generally not all considered in resource assessment methodologies. Our goal is to use these factors as the basis for additional criteria for conducting resource assessments. The resource that satisfies this more complete set of criteria, here termed the "viable" resource, more accurately approximates that which has a reasonable likelihood of actually being developed and produced.

To improve the ability of policymakers to understand how much resources are available and at what cost and impact, the Hewlett Foundation asked RAND to:

- review the existing resource assessment methodologies and results

- evaluate recent studies addressing federal land access restrictions in the Intermountain West

- consider a set of criteria that can be used to define the viable resource, with particular attention on issues relevant to the Intermountain West

- develop a more comprehensive assessment methodology for the viable resource

- employ this methodology to assess the viable resource in Intermountain West basins.

This is an interim report that focuses primarily on the first three points.

Approach

The premise of our study is that it is the viable resource, rather than the traditionally used technically recoverable resource, that is relevant to policymakers regarding their decisions on gas and oil exploration and development. The factors that determine viability include technical recoverability, wellhead economics, infrastructure economics, and environmental acceptability. The effect of these criteria on the available resource is shown conceptually in Figure 1.1. Application of each criterion successively reduces the

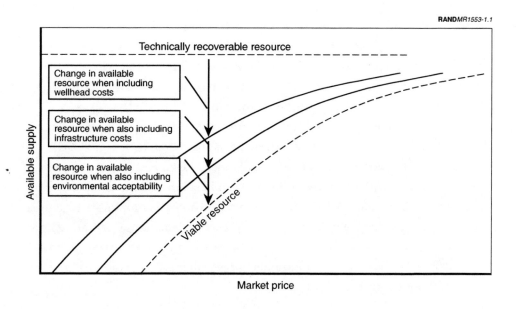

RAND*MR1553-1.1*

Figure 1.1—Effect of Viability Criteria on the Available Resource

amount of resource available at a given price. The axes in Figure 1.1 are purposely reversed relative to conventional economic supply curves. This choice of using supply as the dependent variable emphasizes the fact that the amount of resource available depends on economic and environmental considerations. Note that the curve for environmental acceptability is conceptual only—we do not propose to calculate environmental costs. Rather, we intend to estimate the amount of the economically recoverable resource that can be extracted within a given level of environmental impact. All of these criteria reflect well-known and often cited issues that determine the availability of gas and oil resources. However, though the issues are familiar, little progress has been made toward including them in traditional resource assessments.

This report is organized around these different factors. Each section reviews existing studies and results, evaluates the traditional methods, and proposes alternative approaches. Section 2 presents a critical review of existing hydrocarbon resource assessments. This review is important for understanding the basic categories of gas and oil resources and the historical approaches used to estimate their abundance. These estimates of technically recoverable resources form the foundation upon which additional criteria can be applied. We review the traditional methodologies and highlight the key assumptions. We discuss the results in the context of the differing methods and assumptions made in each assessment. Evaluations are presented for both the lower-48 states and the Intermountain West.

Section 3 evaluates two important studies addressing legal access to gas resources in the Intermountain West. Although we will not include legal access among the criteria for our definition of the viable resource, recent debate about access to Intermountain West resources centers largely on conclusions made in these reports. As a result, it is important to evaluate these studies. One, conducted as part of the recent National Petroleum Council study (National Petroleum Council, 1999), addresses the entire Rocky Mountain Region. The other, led by the Department of Energy (Advanced Resources International, 2001), focuses on the Greater Green River Basin in southwestern Wyoming and northwestern Colorado. We examine the methods and assumptions used in each study and summarize the results. We then present calculations illustrating how some assumptions affect the impact of land access restrictions on the available gas resource. We also identify additional improvements that could be made to better evaluate the restricted resource.

Section 4 discusses existing approaches to incorporate economic considerations into resource assessments. It focuses on the approaches used by the U.S. Geological Survey (Attanasi, 1998) and in the Hydrocarbon Supply Model (Vidas et al., 1993) used by the National Petroleum Council. The method used by both is to balance exploration and production costs with expected resource revenues to determine if it is economically logical to proceed. After reviewing the methods and results, we propose several improvements aimed at tailoring the methods to account for costs specific to the Intermountain West.

Section 5 introduces the concept that resources must be supported by sufficient infrastructure in order to be viable. The impetus for this criterion is that some of the otherwise viable resource in the Intermountain West cannot be developed without substantial increases in the existing pipeline and road infrastructure. For resources to satisfy this criterion, they must either be located close to existing infrastructure or be economically recoverable even after all the necessary infrastructure augmentation costs are accounted for.

Section 6 presents an approach to incorporating environmental impacts into resource assessments. We introduce this criterion to consider potential changes in environmental conditions that could result from gas and oil exploration and production. After discussing the potential impacts of the different steps in the industrial process, we present a framework for considering overall environmental impacts. The goal of this approach is to provide policymakers with a tool to evaluate assumptions about varying environmental impacts that could occur from gas and oil development, and to provide a framework with which to identify areas for exploration and development in the Intermountain West based upon assumed acceptable levels of environmental impact. Section 7 presents our conclusions.

2. Technically Recoverable Gas and Oil Resources

It is impossible to measure the precise volume of natural gas, oil, or natural gas liquid in deposits under the surface of the earth. Many deposits remain undiscovered or unexplored. Different methodologies use a combination of physical evidence (historical production trends, drilling data, seismic information) and statistical methods to estimate the volumes of resources. Numerous organizations conduct resource assessments with varying degrees of complexity. For purposes of this analysis, we compare four assessments conducted by the National Petroleum Council (National Petroleum Council, 1999), the U.S. Geological Survey (U.S. Geological Survey National Oil and Gas Resource Assessment Team, 1995), the U.S. Minerals Management Service (Minerals Management Service, 2000), and the Potential Gas Committee (Potential Gas Committee, 2001).

The goal of these assessments is to estimate the potential supply of natural gas and oil resources, which, combined with estimates of the proved reserves, make it possible to appraise the nation's long-range gas and oil supply. The assessments do not necessarily reflect the amount expected to be actually recovered. For example, in the Potential Gas Committee assessment, "No consideration is given whether or not this resource will be developed; rather, the estimates are of resources that could be developed if the need and economic incentive exist." Similarly, the U.S. Geological Survey assessment "makes no attempt to predict at what time or what part of potential additions will be added to reserves. For the National Assessment, resources and potential reserve additions are evaluated regardless of political, economic, and other considerations."

These assessments estimate the "technically recoverable" resource, or the amount judged to be recoverable given certain assumptions about technical capabilities. In practice, the definition of the term "technically recoverable" is unclear and is inconsistently applied among the different assessments. A large part of the difference between existing resource assessments results from differing assumptions as to what constitutes a technically recoverable resource.

This section discusses the different methodologies and assumptions made in each of the four assessments and consequently the different results, both for the U.S. lower 48 states in general and for the Rocky Mountain Region in particular. The section reveals that differences between assessments arise from a variety of different sources, including varying interpretations of terms, differing assumptions, and differences in the status of resource exploration, particularly in the Intermountain West, at the different dates the assessments were conducted.

Different Definitions of Terms

At the outset, the existing resource assessments use different definitions of the terms used to describe a resource. There is no universally accepted set of definitions for these terms. For clarity, Figure 2.1 presents a hierarchy of terms used to describe typical resource categories. The reliability of resource assessments in each of the different categories decreases from left to right in the figure as resource abundances become more speculative. The four assessments cited above define many of these categories differently; details of these differences are listed in Table A.2 in the Appendix. We use general terms to describe each category below the figure.

RAND*MR1553-2.1*

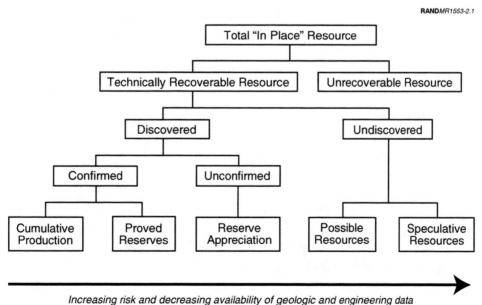

Increasing risk and decreasing availability of geologic and engineering data

SOURCE: Modified from Potential Gas Committee (2001).

Figure 2.1—Hierarchy of Categories to Describe Resources

8

Total "In-Place" Resource: The finite volume of the resource that exists on earth before any production occurs.[1]

Technically Recoverable Resource: The portion of the total in-place resource that can potentially be recovered given current or anticipated technology and a qualitative consideration of expected economic conditions.[2] This is the estimate presented in most resource assessments.

Unrecoverable Resource: A major portion of the total in-place resource is unrecoverable, including resources scattered in small deposits or in locations where the resource is virtually impossible to recover with current or anticipated technology.

Discovered Resource: The technically recoverable resource includes both discovered and undiscovered resources. The discovered resource includes both that which has been *confirmed* (cumulative production and proved reserves) and *unconfirmed* (reserve appreciation).

Undiscovered Resource: The resource that may be discovered in new fields in provinces that are currently productive or unproductive.

Cumulative Production: The historical cumulative volume of resources withdrawn from producing reservoirs. Reliable historical records of U.S. production date back to the 1930s, which are estimated to account for 95 percent or more of total production to date.

Proved Reserves: Estimated quantities of a resource that current analysis and geologic and engineering data demonstrate with reasonable certainty to be recoverable in the future from known reservoirs under existing economic and operating conditions. These resources are deemed to be confirmed, because they are associated with producing reservoirs or have been extensively tested.

Reserve Appreciation: The resource expected to result from future extensions in existing pools in known producing reservoirs. The resource is unconfirmed, because, although they have been discovered, the extent of the pools have not been completely defined.

---

[1]Despite the name, estimates of the total in-place resource typically exclude more exotic resources such as gas hydrates, geopressured-geothermal accumulations, and deep-earth gas.

[2]This economic consideration is ill-defined and of little relevance. Technically recoverable resources are not subject to a rigorous economic analysis such as that described in Section 4 of this report.

Possible Resources: Undiscovered resources that exist outside known producing fields but that are associated with productive formations in producing provinces. Their existence is postulated by the projection of plays into less explored areas of the same province under both similar and different geologic conditions.

Speculative Resources: Undiscovered resources in formations or geological provinces that have not yet proven productive.

In addition, each of these categories can contain *conventional* and *nonconventional* resources. Conventional resources are typified by downdip water contacts and can be extracted using traditional development practices. Nonconventional resources, sometimes referred to as continuous deposits, include resources contained in low-permeability sandstone ("tight sandstone" or "tight gas"), shale, chalk, and coalbed methane.

Assessment Methodologies

Each assessment uses a different methodology to estimate reserve appreciation, undiscovered resources, and nonconventional resources. In general, the U.S. Geological Survey and U.S. Minerals Management Service assessments were done at the play level using various simulation techniques, whereas the National Petroleum Council and Potential Gas Committee were done at the region or basin level, relying on individual estimators and expert panels using statistical methods. An important distinction between the different methodologies is their assumption regarding technology impacts. The U.S. Geological Survey assesses only resources available using current recovery technology, while the National Petroleum Council and Potential Gas Committee assessments assume a certain rate of technology advancement. Further details of the specifications used in each assessment, such as effective dates, commodities and resource categories assessed, and areas covered, are listed in the Appendix.

Reserve Appreciation

The resource in a producing field is proved over a period of years or decades. Only a portion of the particular resource in a field will be proved and available for production in a given year. Estimated Ultimate Recovery (EUR) is the term used to describe the total resource in a field, or the sum of cumulative production plus proved reserves at any given date. In general, EUR increases over time. The difference between the current EUR and ultimate total production from a field is reserve appreciation.

The National Petroleum Council estimated reserve appreciation using a statistical approach that assumes that successive drilling produces declining additions to EUR. In other words, the largest resource pockets in a field are typically targeted and found first, and then the producer moves onto successively smaller pockets. By extrapolating the decline in recoveries, it is possible to estimate the total reserves and, by subtraction of proved reserves, the reserve appreciation.

The U.S. Geological Survey and U.S. Minerals Management Service, in contrast, relied on growth functions that related total field size (cumulative production plus proved reserves) to field age. Field age is used as a proxy for the degree of field development. The key assumptions here are that (1) the amount of growth in any one year is proportional to the size of the field and that (2) this proportionality varies inversely with the age of the field. To estimate reserve appreciation, the growth functions projected growth out 80 years.

The Potential Gas Committee divides the potential resource into three categories: probable, possible, and speculative. The probable category represents the further development of fields that have already been discovered, including extensions and new pool discoveries. This category is equivalent to reserve appreciation.[3] An expert panel derived the estimates based on comparative factors of known resources, either in the same geological province or in similar provinces. The committee's documentation states, "In its simplest form, the estimate of the potential gas supply is derived by (1) estimating the volume of potential gas-bearing reservoir rock, (2) multiplying this volume by a yield factor, and (3) discounting to allow for the probability that traps and/or accumulations exist." Under this approach, the judgments and experience of the individuals are central to the credibility of the estimate.

Undiscovered Conventional Resource

The National Petroleum Council estimate of undiscovered conventional resource was derived through a consensus update of a resource assessment completed as part of an earlier natural gas study (National Petroleum Council, 1992). The 1999 assessment update was based on discussions among people from industry, government, and associations, published information dating back to 1992 (including the 1995 U.S. Geological Survey assessment), discovery and

---

[3]Confirmed through personal communication with John Curtis, director of the Potential Gas Agency (the body charged with completing the Potential Gas Committee estimate).

production trends between 1992 and 1999, and unpublished company resource estimates. The 1999 assessment did not involve extensive modeling or analysis, but did include analysis of discovery trends by basin and depth interval and a complete review of the deep-water Gulf of Mexico. The resource estimates were revised where current industry expectations differed from the 1992 assessment.

In contrast, the U.S. Geological Survey and U.S. Minerals Management Service based their assessments on detailed quantitative analyses. For example, the U.S. Geological Survey first estimated the possible size, number, and type of resource accumulations within a geologic play and the associated play risk. Play risk expresses the probability of hydrocarbon occurrence based on charge, reservoir, and trap. Plays were not assessed when the play probability was 10 percent or less. Estimators then employed discovery-processing modeling, reservoir-simulation modeling, play analogs, and spatial analyses. The assessments evaluated both confirmed and unconfirmed (or hypothetical) deposits.

The Potential Gas Committee includes both possible and speculative resources in its estimate of undiscovered resources. The general methodology used here was identical to that described for reserve appreciation: An expert panel derived its estimates based on comparative factors of known resources, either in the same geological province or in similar provinces.

Nonconventional Resources

Nonconventional resources include tight sandstone, chalk, shale, coalbed methane, and low-Btu gas. Among these, typically only the potential formations that are well known and have had some exploration or development activity are included in resource assessments.

The National Petroleum Council nonconventional resource assessment covered tight sandstone, shale, and coalbed methane. As was the case with undiscovered resources, the 1999 assessment was largely a consensus update of the 1992 study. In the 1992 study, the nonconventional gas assessment was completed using a consensus approach in which subgroups examined each type of nonconventional resource. Data for the 1992 study were collected through a variety of means: for tight sandstone, a confidential survey of operators in known formations; for shale, 1980s data combined with subsequent company field experience; and for coalbed methane, data from 20 coalbed methane basins, 8 of which included detailed data. For the 1999 assessments, the study participants reviewed the earlier estimates and compared them with actual discovery and production trends since 1992. The existing National Petroleum Council assessments of coal-bed methane were also compared with the 1995 U.S. Geological Survey

12

assessments and, in some cases, the USGS assessments were used. Adjustments were made where it was felt the resource estimates were inconsistent with more recent trends. However, no detailed survey of operators or other analytical method was used for the 1999 study.

The U.S. Geological Survey defined two types of nonconventional deposits: (1) tight sandstone, shale, and chalk and (2) coalbed methane. The methodology to assess tight sandstone, shale, and chalk followed a four-step process of mathematical modeling that incorporated no explicit consideration of technological improvements or economics. Future expectations were based on historical production and modeling patterns. The methodology to assess coalbed methane was similar, but it also relied heavily on production forecasting using a reservoir simulator.

The U.S. Minerals Management Service did not consider nonconventional resources in federal offshore areas.

The Potential Gas Committee assessed only coalbed methane among the nonconventional resources. Its assessment nominally includes some tight sandstone and shale gas with its conventional resource estimate, but they are not inventoried separately and thus cannot be compared with the other assessments. The methodology used to estimate the coalbed methane resource was very similar to the technique used by the committee for reserve appreciation and undiscovered conventional resources, estimating the recoverable resource based on assumptions about a range of recovery factors. Once again, the knowledge and experience of the estimators were key to developing the assessment.

Results for the Lower 48 States

Table 2.1 compares the assessments of the technically recoverable resource for the U.S. lower 48 states. The results for the main categories for natural gas are displayed in Figure 2.2. In terms of conventional resources, the National Petroleum Council's estimate is the largest; the combined U.S. Geological Survey and U.S. Minerals Management Service assessment is the smallest. In terms of nonconventional resources, the Potential Gas Committee appears pessimistic, because it includes the fewest resource categories.

The variations between the assessments result from differences in insight into the size and distribution of the resource base, assumptions about the effect of current

Table 2.1

Comparison of Lower-48 Technically Recoverable Resource Assessments (as of December 31, 2000)[a]

| | Natural Gas (tcf) | | | Oil (BB) |
	NPC	USGS/MMS	PGC	USGS/MMS
Conventional resources				
Reserve appreciation[b]	272	291	170[i]	59
New fields[c]	620	408	572[i]	77
Subtotal	892	699	742	136
Nonconventional resources				
Tight sandstone	218	280[f]	–[g]	}2
Devonian shale	50	–[g]	–[g]	
Coalbed methane	70	45	98	
Other	13	–[h]	–[h]	
Subtotal	351[d]	325[e]	98	2
Total potential resources	1,243	1,014	840	138
Proved reserves	168	168	168	21
Total future supply	1,411	1,182	1,008	159
Cumulative production	936	936	936	172
Total ultimately recoverable	2,347	2,118	1,944	331

[a]Each of the resource assessments was dated using a different end point. The NPC, USGS, MMS, and PGC assessments were dated 1/1/1998, 1/1/1994, 1/1/1999, and 12/31/2000, respectively. All of the assessments have been adjusted to reflect an end date of 12/31/2000.

[b]Reserve appreciation resources are assumed to account for 50 percent of gas and 70 percent of oil production plus transfers to proved reserves between assessment end point and 12/31/2000. Reported reserve appreciation values were reduced by 33 tcf, 61 tcf, and 7.8 BB for NPC gas, USGS/MMS gas, and USGS/MMS oil, respectively, to reflect this assumption.

[c]New field resources are assumed to account for 20 percent of gas and 30 percent of oil production plus transfers to proved reserves between assessment end point and 12/31/2000. Reported new field resource values were reduced by 13 tcf, 25 tcf, and 3.4 BB for NPC gas, USGS/MMS gas, and USGS/MMS oil, respectively, to reflect this assumption.

[d]Nonconventional resources are assumed to account for 30 percent of gas production plus transfers to proved reserves between NPC end point and 12/31/2000. Reported nonconventional resource values were reduced by 20 tcf to reflect this assumption. This was distributed as 12.6 tcf to tight sandstone, 4 tcf to coalbed methane, 2.6 tcf to shale, and 0.8 tcf to Other.

[e]Nonconventional resources are assumed to account for 30 percent of gas production plus transfers to proved reserves between USGS end point and 12/31/2000. Reported nonconventional resource values were reduced by 37 tcf to reflect this assumption. This was distributed as 31.9 tcf to tight sandstone and 5.1 tcf to coalbed methane.

[f]Includes tight sandstone, shale, and chalk.

[g]Not reported as a separate category.

[h]Not assessed.

[i]Includes tight sandstone and shale.

NOTE: tcf = trillion cubic feet, BB = billion barrels.

and future technology on resource recovery, the estimator's judgment, and variations in the resource categories included in each assessment. Under conventional resources, for example, the Potential Gas Committee estimates the

14

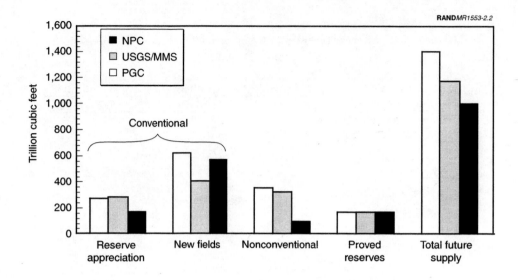

Figure 2.2—Comparison of Lower-48 Technically Recoverable Natural Gas Resource Assessments

lowest potential for reserve appreciation, but relatively high potential for undiscovered conventional resources. These results reflect differing assumptions about the remaining resource in existing fields and the impact of new technology.

In contrast, the low estimate for the new field resource from the U.S. Geological Survey and U.S. Minerals Management Service assumes that resources are generally more mature. In addition, the U.S. Geological Survey assessment emphasizes established plays and places little resource in unconfirmed plays. The assessment also reflects the less advanced state of drilling technology at the time the assessment was conducted. Onshore drilling technology has improved substantially since the U.S. Geological Survey completed its onshore assessment in 1995.

The National Petroleum Council assessment, being the most recent, shows the highest amount of coalbed methane resource. Spurred by federal tax credits enacted in the early 1980s, development of coalbed methane has been increasing rapidly in the past two decades. The higher coalbed methane estimate in the National Petroleum Council assessment simply reflects the newness of information about this particular resource.

Results for the Rocky Mountain Region

The primary focus of this analysis is the Rocky Mountain Region. For the
purposes of comparison between assessments, we use the National Petroleum
Council's definition for the Rocky Mountain Region, which includes their
Williston Basin, Overthrust Belt, Rocky Mountain Foreland, and San Juan Basin
assessment provinces, as shown in Figure 2.3. These provinces include the entire
states of Colorado, Wyoming, Montana, and North Dakota and major portions of
Arizona, New Mexico, Utah, South Dakota, and Nebraska. The U.S. Geological
Survey and Potential Gas Committee delineate a larger number of assessment
provinces, allowing us to select provinces to closely match the National
Petroleum Council assessment regions.

There are a large number of plays within the area experiencing varying levels of
activity. The greatest resource potential in the Rocky Mountain Region is
concentrated in the Uinta/Piceance Basin in northwestern Colorado and
northeastern Utah, the Greater Green River Basin in southern Wyoming, the
Powder River Basin in northern Wyoming, the San Juan Basin in northwestern

RANDMR1553-2.3

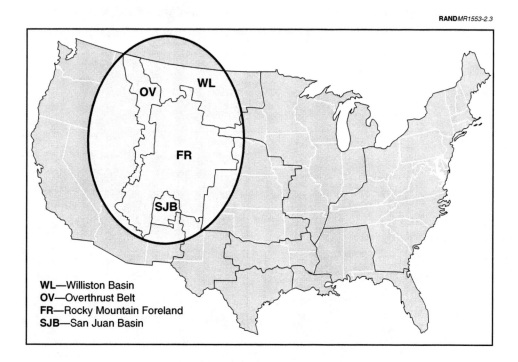

Figure 2.3—Rocky Mountain Resource Regions

Mexico, the Wind River Basin in central Wyoming, and the Montana Folded Belt in western Montana.

Table 2.2 compares the Rocky Mountain Region resource assessments for the four geologic provinces shown in Figure 2.3. This table excludes the federal offshore assessment from the U.S. Minerals Management Service, which is irrelevant in the Rocky Mountain Region. The results for the main categories for natural gas are displayed in Figure 2.4.

The focus of our comparison is the potential gas resource. Once again, the National Petroleum Council is the most optimistic regarding both conventional and total resources. The U.S. Geological Survey estimates the least conventional resources but greatest nonconventional resources. Again, the Potential Gas Committee estimate for nonconventional resources includes only coalbed methane.

Nonconventional resources are particularly important in the Rocky Mountain Region. As Figures 2.5 and 2.6 show, coalbed methane and tight sandstone constitute a major portion of the total resource in the Rocky Mountain Region. Tight sandstone, in fact, accounts for 37 to 65 percent of the total resource in the region (Table 2.2), so excluding it distorts any comparison of the total potential resource.

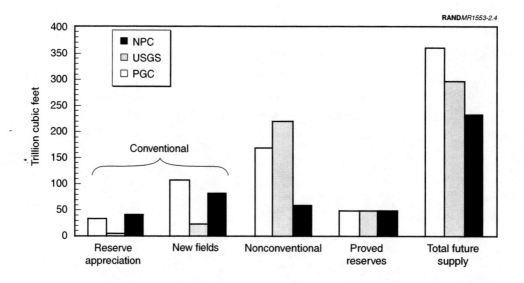

Figure 2.4—Comparison of Rocky Mountain Region Technically Recoverable Natural Gas Resource Assessments

Table 2.2

Comparison of Rocky Mountain Region Technically Recoverable Resource Assessments (as of December 31, 2000)[a]

	Natural Gas (tcf)			Oil (BB)
	NPC	USGS	PGC	USGS
Conventional resources				
Reserve appreciation[b]	34	6	41[j]	10.6
New fields[c]	107	24	83[j]	4.9
Subtotal	141	30	124	15.5
Nonconventional resources				
Tight sandstone	132	195[f]	–g	}1.0
Devonshire shale	0	–g	–g	
Coalbed methane	38	24	59	
Other	0	–h	–h	
Subtotal	170[d]	219[e]	59	1.0
Total potential resources	311	249	183	16.5
Proved reserves	49	49	49	2.3
Total future supply	360	298	232	18.8
Cumulative production[i]	99	99	99	NA
Total ultimately recoverable	459	397	331	NA

[a]Each of the resource assessments was dated using a different end point. The NPC, USGS, and PGC assessments were dated 1/1/1998, 1/1/1994, and 12/31/2000, respectively. All of the assessments have been adjusted to reflect an end date of 12/31/2000.

[b]Reserve appreciation resources are assumed to account for 50 percent of gas and 70 percent of oil production plus transfers to proved reserves between assessment end point and 12/31/2000. Reported reserve appreciation values were reduced by 10 tcf, 16 tcf, and 1.2 BB for NPC gas, USGS gas, and USGS oil, respectively, to reflect this assumption.

[c]New field resources are assumed to account for 20 percent of gas and 30 percent of oil production plus transfers to proved reserves between assessment end point and 12/31/2000. Reported new field resource values were reduced by 4 tcf, 6.6 tcf, and 0.5 BB for NPC gas, USGS gas, and USGS oil, respectively, to reflect this assumption.

[d]Nonconventional resources are assumed to account for 30 percent of gas production plus transfers to proved reserves between NPC end point and 12/31/2000. The nonconventional resources were reduced by 6.0 tcf to reflect this assumption. This was distributed as 4.6 tcf to tight sandstone and 1.4 tcf to coalbed methane. Data for shale and other were not available on a regional basis.

[e]Nonconventional resources are assumed to account for 30 percent of gas production plus transfers to proved reserves between USGS end point and 12/31/2000. The nonconventional resources were reduced by 9.7 tcf to reflect this assumption. This was distributed as 8.6 tcf to tight sandstone and 1.1 tcf to coalbed methane.

[f]Includes tight sandstone, shale, and chalk.

[g]Not reported as a separate category.

[h]Not assessed.

[i]RAND estimate.

[j]Includes tight sandstone and shale.

NOTE: tcf = trillion cubic feet, BB = billion barrels, NA = not available.

The U.S. Geological Survey assessment for conventional resources is quite low and may be inconsistent with recent production rates. Between 1994, the date of the U.S. Geological Survey assessment, and 2000 there was about 20 trillion cubic

18

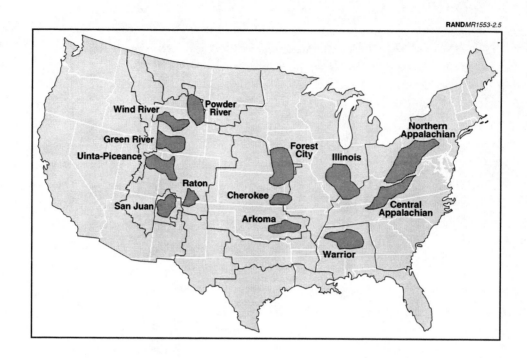

Figure 2.5—Basins Containing Coalbed Methane Deposits

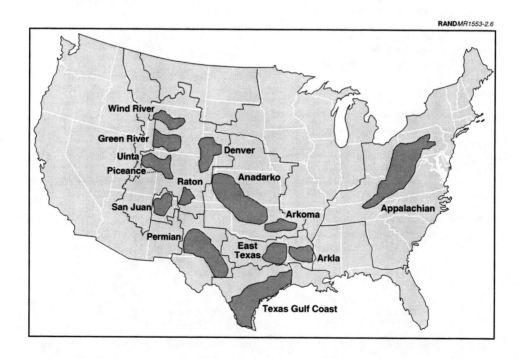

Figure 2.6—Basins Containing Tight Sandstone Gas Deposits

feet (tcf) of natural gas production in the Rocky Mountain Region, with production rates increasing substantially since 1997 (Energy Information Administration, 2001). While much of the recent focus of activity in the Rockies is on coalbed methane, there is still substantial conventional resource activity. This high and increasing production activity suggests that substantial resource remains and that the U.S. Geological Survey conventional resource estimate, while consistent with assumptions at the time it was made, may not reflect recent discovery trends, reserve appreciation, and technical advances.

The case for coalbed methane is similar. The U.S. Geological Survey estimate is substantially lower than the others. Initially spurred by section 29 federal tax credits and intensified by relatively higher gas prices in recent years, there has been a great deal of coalbed methane exploration and development activity in the Rocky Mountain Region. Over this time, the assessments of the resource have generally expanded. Preliminary results from the new U.S. Geological Survey assessment, for example, show a 14-fold increase in the coalbed methane resource in the Powder River Basin (Energy Information Administration, 2002).

The higher value of the tight sandstone resource in the U.S. Geological Survey assessment appears to reflect its earlier effective date. As more has been learned from actual exploration and production activity in recent years, some resource assessments have gradually reduced their expectations for the tight sandstone resource.

Gas production in the Rocky Mountain Region grew from 1.4 tcf in 1986 to almost 3.7 tcf by 2000, accounting for over 80 percent of the total growth in lower-48 natural gas production during this period. The growth occurred despite lease restrictions, difficult terrain, and constraints on take-away pipeline capacity in the region.

Although the assessments vary, they suggest that the Rocky Mountain Region has the potential to remain a major contributor to U.S. energy supply for many years to come. However, the very high proportion of nonconventional resources in the region introduces a higher than normal degree of uncertainty on the amount of that resource that is ultimately recoverable.

Resource Assessment Evolution

It is important to reiterate that resource assessments are highly uncertain. By definition, assessments exclude major portions of potential resources because of assumptions about technological improvements, economics, and other factors. By and large, assessments only consider those resources that are familiar to the

20

energy industry. That does not mean that the excluded resources are necessarily unknown, only that an implicit assumption is made that they will not be recoverable in the foreseeable future. However, as illustrated in Figure 2.7 (from Holtberg, 2000), resource assessments have grown with time. This increase results primarily from two causes: increased understanding of the resource (gained from new exploration) and increased ability to recover the resource (resulting from technological improvements).

For example, the deep waters of the Gulf of Mexico were often called a "dead sea" during the early to mid-1980s, because it was felt that the resource was inaccessible. As a result, the deep-water resource was often totally excluded or underassessed in resource assessments. With advances in technology, the Gulf is now one of the fastest growing areas of exploration and production. Not only is the resource in the Gulf now assessed, but the estimates of the size of the resource have grown rapidly over the last decade. Another example is the coalbed methane resource in the Rocky Mountain Region. The resource was known to exist for decades, but an implicit assumption was made that it would never be recoverable due to technological and economic limits. In 2000, roughly 1.3 tcf of natural gas was produced from coalbed methane basins in the Rocky Mountain Region (up from about 0.5 tcf in 1985). Coalbed methane is now

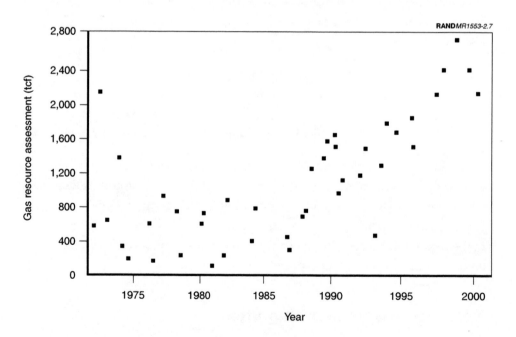

Figure 2.7—Historical Gas Resource Estimates

included in resource assessments, and the amount assessed is rapidly being revised upward as experience grows.

3. Legal Access To Resources in the Intermountain West

Much of the current debate over resource development in the Intermountain West stems from restrictions on federal land use and their implications for accessing gas and oil resources. While interest in access restrictions has existed for some time, recent efforts to evaluate these restrictions and their impact have been spurred largely by the Energy Policy and Conservation Act of 2000, which directs federal land management agencies to assess the energy potential of public lands and identify impediments to its development. As a result, considerable effort is being expended on quantifying the amount of gas and oil resources underlying federal lands that is subject to various forms of access restrictions. Two important studies addressing gas resource access on federal lands in the Intermountain West have been completed recently. One, conducted as a part of the National Petroleum Council's 1999 natural gas report (National Petroleum Council, 1999), addressed the Rocky Mountain Region, while the other, prepared for a multi-agency audience led by the Department of Energy (Advanced Resources International, Inc., 2001), focused on the Greater Green River Basin.

These studies conclude that substantial amounts of gas resources in the Intermountain West are inaccessible or accessible with restrictions as a result of various types of federal lands access restrictions. These restrictions, formally known as lease stipulations, are conditions accompanying a lease, usually for environmental protection reasons (but also for historical and cultural reasons), that dictate where, how, and when drilling activities may be conducted. The results of these studies have led industry to call for reduced restrictions and to "continue the work begun with this [National Petroleum Council] study to inventory existing information on the resource base in the Rocky Mountains and analyze the impact of access restrictions." Work is currently under way to assess access restrictions in a number of Rocky Mountain basins.

In their effort to identify impediments to energy development, these studies make some important assumptions that have implications for the impact of federal land access restrictions on the accessibility of gas resources. Below, we identify areas in which the approach could be enhanced and illustrate how some simple modifications in the approach influence the outcome of the analysis.

While efforts to quantify resources subject to access restrictions may be important, they bypass the more fundamental question of how much viable resource is present in the areas being considered. It is of limited value to estimate the amount of technically recoverable resource that is subject to legal access restrictions because much of this resource will remain inaccessible for other reasons. Nonetheless, because of the debate prompted by these studies, the ongoing efforts to conduct this type of analysis throughout the Intermountain West, and the potential for the results of these efforts to influence policy decisions regarding access to federal lands, it is important to evaluate this work.

Methodology

The methodology used in the Rocky Mountain and Greater Green River Basin studies is similar. The first step entails collecting lease stipulation information from the agencies with jurisdiction over the federal lands in the study area. Over 80 percent of the federal land in the Rockies is managed by the Bureau of Land Management, the Forest Service, and the Bureau of Indian Affairs. The remaining land is split among a number of classifications and agencies, including Wilderness Areas, the National Park Service, the Fish and Wildlife Service, National Recreation Areas, and several others. Geographic Information System (GIS) files of federal and Indian lands, obtained from the Bureau of Land Management, were used to inventory the acreage within the study areas. The access status of federal lands was then identified from maps showing the environmental stipulation areas as well as descriptions of the stipulations in each of the areas.

The Greater Green River Basin study lists 108 different lease stipulations collected from Bureau of Land Management and Forest Service offices in the study area. The stipulations specify the terms of protection of land attributes (e.g., cannot disturb elk calving). With the help of federal agencies, analysts interpreted the area that the stipulations would cover and the effects of the stipulations on access for drilling. Table 3.1 lists the stipulation categories and their effects on gas drilling, as used in both the Rocky Mountain and Greater Green River Basin studies, as well as the access levels that the National Petroleum Council assigned to the different stipulation categories in the Rocky Mountain study.

Both the Rocky Mountain and Greater Green River Basin studies use townships (six miles by six miles) as the basic unit of land measurement. Where multiple stipulations apply to one location, the most restrictive was assigned to that location. Where multiple timing limitation stipulations specifying different parts of the year apply to a given location, the location was assigned the cumulative time limitation of all applicable stipulations. In the Greater Green River Basin

24

Table 3.1

Classification of Lease Stipulations and Effect on Gas Drilling

Stipulation Category	Effect on Drilling	Rocky Mtn. Access Level
No Access (statutory)	No drilling	
No Access (administrative) (includes No Leasing and No Surface Occupancy)	No drilling	
Timing limitations > 9 months/year	Precludes drilling during portions of the year	No Access
6 - 9 months/year 3 - 6 months/year		High Cost
< 3 months/year		Standard Lease Terms
Controlled Surface Usage	Varied, may be mitigated	High Cost
Combinations of Controlled Surface Usage and Timing Limitations		
Standard Lease Terms	No restrictions	Standard Lease Terms

SOURCES: National Petroleum Council (1999); Advanced Resources International 2001).

study, the resources within a township are allocated in proportion to the area covered by each of the stipulation categories.

In the Rocky Mountain study, the lease stipulation categories were estimated by extrapolating the results of analyses of six "calibration areas." The calibration areas consisted of three Bureau of Land Management districts (Pinedale, Price, Rock Springs) and three Forest Service districts (Bridger-Teton, Manti-La Sal, Uinta). These areas total 14.8 million acres of federal land, or 10 percent of the federal land in the Rockies, and contain about 30 percent of the gas resources in the study area. These calibration areas were chosen because of their high resource levels and industry activity.

Lease stipulation categories for the Bureau of Land Management and Forest Service lands in the complete Rocky Mountain Region study area were assigned by applying the apportionment determined from the calibration areas. Of the remaining federal lands in the region, the lands of the National Park Service, Fish and Wildlife Service, National Recreation Areas, and Wilderness Areas were classified as No Access. Apportionment of lease stipulation categories to Bureau of Reclamation, Department of Defense, Department of Energy, Indian, and other federal lands was made by educated guesses by the National Petroleum Council Policy Group.

The final step was to estimate the impact of the lease stipulations on the amount of resource available. For the National Petroleum Council Rocky Mountain study, gas resources were assigned to one of three access levels: No Access, High Cost, and Standard Lease Terms (Table 3.1). Resources in areas subject to stipulations that restrict drilling for nine months per year or more were classified as No Access based on the typical drilling times required in the majority of areas. Similarly, areas subject to stipulations that restrict drilling for three months per year or less were classified as Standard Lease Terms. Areas subject to stipulations that restrict drilling from three to nine months per year were classified as a "gray area" in which resources are available, but only with a penalty of higher cost and delayed development.

The Greater Green River Basin study did not assign access levels. Results were presented simply as the amount of gas resources underlying federal lands in each of the stipulation categories shown in Table 3.1. However, Advanced Resources International also outlined a more complex method for estimating the impact of the lease stipulations on resource access in the Greater Green River Basin based on drilling opportunities. For each township, the time required to drill a well was compared with the time allowed to drill to determine whether drilling could proceed. The analysts first assumed that all wells would be drilled to a depth half-way to the Precambrian basement rocks (this was intended to approximate the depth to the lower Cretaceous section, the location of much of the tight sandstone). Then the analysts presented drilling depth versus time relationships for wells of 10,000, 14,000, and 18,000 feet in depth. Based on these relationships (and assuming that drilling must be completed in one season), wells that are 14,000 feet deep or less are precluded only in locations in which cumulative Timing Limitations amount to more than nine months per year. Wells deeper than 14,000 feet are precluded in areas restricted for more than six months per year. Using GIS to correlate the locations of gas plays, Timing Limitations, and the depth to the basement (used to estimate well depths), the analysts then indicated for each township whether or not industry would be able to drill at least one well in a season.

The Greater Green River Basin study does not explicitly report the amount of resources underlying townships in which drilling is precluded. It does, however, present a map showing the drilling opportunity status (possible or precluded) of each township, as well as the resource amount under each township. From these we were able to estimate the amount of restricted resource, which we present below.

Results

The results of the access restriction studies are presented in Tables 3.2–3.4. Table 3.2 shows the unproved gas resources in the Rocky Mountain Region as a function of access level. The results show that approximately 60 percent of the unproved gas is available under Standard Lease Terms, with the remaining 40 percent available at High Cost or Inaccessible.

Table 3.3 shows the amount of undiscovered conventional and nonconventional gas resources underlying federal lands in the Greater Green River Basin in each lease stipulation category. Compared to the entire Rockies, more resource is subject to access restrictions in the Greater Green River Basin, with 29.5 percent closed to development and 38.5 percent available with restrictions; 32 percent is available under Standard Lease Terms.

Based on the previously described drilling opportunities approach, however, we find that 66 percent of the gas in the Greater Green River Basin is accessible (Table 3.4). This amount is only slightly less than the combined total of gas subject to Standard Lease Terms and available with restrictions (70.5 percent, Table 3.3). This result indicates that nearly all of the gas in the Greater Green River Basin in areas subject to Timing Limitations and Controlled Surface Usage is accessible for production using standard drilling operations. Given the rather simple assumptions regarding drilling time and drilling depth in the drilling opportunities analysis, the exact amount of accessible gas is highly uncertain. Nonetheless, this exercise demonstrates that land access restrictions do not necessarily prohibit resource extraction, even when using standard-cost, single-season drilling techniques.

The results summarized in Tables 3.2–3.4 indicate that moderate to substantial fractions of the gas in the Rockies are subject to access restrictions. However, care should be used in interpreting these studies. For example, the access restrictions in the Greater Green River Basin shown in Table 3.3 represent only the resources underlying federal lands. Non-federal lands are generally considered accessible to industry. This distinction has generated confusion about the results of this study. Using the total land as a basis would reduce the fraction of resources subject to federal lease stipulations and associated access restrictions.

Table 3.2

**Reported Natural Gas Access Restrictions in the
Rocky Mountain Region**

	Unproved Gas Resources[a]	
Access Level	tcf	%
No Access	29.2	8.6
High Cost	108.0	31.7
SLT	203.0	59.7
Total	340.2	100.0

SOURCE: National Petroleum Council
(1999).

[a]While the National Petroleum Council report
text suggests that these results are for gas resources
on federal lands only (National Petroleum Council,
1999, v. II, p. S-20), comparison with the raw data in
Appendix J and discussions with H. Vidas and
B. Hugman indicate that these results reflect both
federal and nonfederal lands.
NOTE: tcf = trillion cubic feet; SLT = Standard Lease
Terms.

Table 3.3

**Reported Natural Gas Access Restrictions in the
Greater Green River Basin**

Access Level	Undiscovered Conventional and Nonconventional Gas on Federal Lands		
	Stipulation Category	tcf	%
Closed to development	No Access (Statutory)	1.4	1.2
	No Access (Administrative) and No Surface Occupancy	33.0	28.3
Subtotal		34.5	29.5
Available with restrictions	TL > 9 mos.	0.50	0.4
	TL 6 to 9 mos.	20.3	17.4
	TL 3 to 6 mos.	21.5	18.4
	TL < 3 mos.	0.86	0.7
	CSU	1.8	1.5
Subtotal		44.9	38.5
SLT		37.4	32.0
Total		116.8	100.0

SOURCE: Advanced Resources International (2001).
NOTE: tcf = trillion cubic feet, TL = Timing Limitations, CSU = Controlled Surface
Usage, SLT = Standard Lease Terms.

Table 3.4

**Natural Gas Drilling Opportunities in the
Greater Green River Basin**

Access Level	Undiscovered Conventional and Nonconventional Gas on Federal Lands	
	tcf	%
Inaccessible (drilling precluded)	40.2	34
Accessible (drilling possible)	76.6	66
Total	116.8	100

SOURCE: Advanced Resources International (2001).
NOTE: tcf = trillion cubic feet.

In addition, it is important to understand the resource basis upon which the access restrictions are applied. As discussed in Section 2, hydrocarbon resources are classified into many different categories, including proved reserves, undiscovered conventional reserves, nonconventional reserves, and reserve appreciation. The Rocky Mountain Region contains substantial amounts of gas in all of these categories. The access restriction results presented in the studies, however, do not include proved reserves or, in the case of the Greater Green River Basin study, reserve appreciation. The effect of excluding proved reserves from the basis is to overestimate the impact of access restrictions because proved reserves are not subject to lease stipulations and hence fall into the Standard Lease Terms category. Excluding reserve appreciation has a similar effect. Reserve appreciation to existing fields is generally subject either to Standard Lease Terms or to some degree of restriction but is generally considered a resource that is available to industry.

To evaluate access restrictions in the context of the total gas resource in the Rocky Mountain Region, we have recast the lease stipulation and access restriction results according to a basis that includes all lands and all gas resources and reserves in the study areas. For the Rocky Mountain Region, Figure 3.1 shows a comparison between the results when including only unproved resources (as listed in Table 3.2) and when adding 35.1 tcf of proved reserves (National Petroleum Council, 1999) to the Standard Lease Terms category. The inclusion of proved reserves results in a relatively small decrease (<4 percent) in the fraction of gas that is inaccessible or available at increase cost.[1]

---

[1] The effect on the distribution of access levels of including proved reserves would be greater when considering only the economically recoverable resources, because proved reserves are by definition all economically recoverable.

RAND*MR1553-3.1*

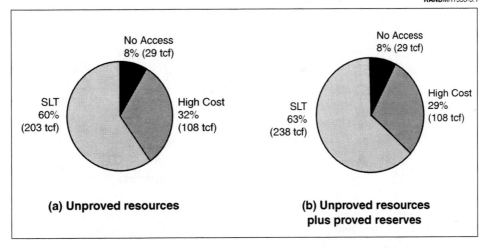

Figure 3.1—Effect of Including Proved Reserves on Access Levels in the Rocky Mountain Region

In the case of the Greater Green River Basin, adjusting the basis has a much larger effect on the relative proportion of access-restricted land. Figure 3.2 shows a comparison between the distribution of access levels when including only undiscovered conventional and nonconventional gas underlying federal lands (as reported in Table 3.3) and when adding proved reserves, reserve appreciation,[2] and gas resources underlying non-federal lands to the Standard Lease Terms category. This adjustment results in a large (24 percent) decrease in the fraction of gas that is inaccessible or available with restrictions.

An additional important issue that is not addressed in the existing studies is that access restrictions have practical impact only on resources that are actually recoverable. This means that, in contrast to the technically recoverable resource estimate, the basis upon which the access restrictions are relevant is the viable resource estimate. The viable resource estimate considers factors in addition to technology and is smaller than the technically recoverable estimate. Thus, when considering the viable resource, the amount of resource that is subject to access restrictions will be less than that reported for the technically recoverable resource.

While the distinction between the technically recoverable and viable resource basis is important when discussing the absolute amount of restricted resource,

_____

[2]Gas from reserve appreciation divided between Restricted Access and SLT (see Figure 3.2 notes).

RAND*MR1553-3.2*

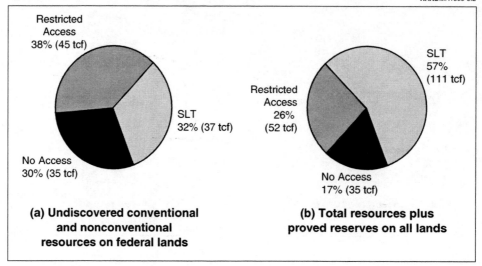

(a) Undiscovered conventional and nonconventional resources on federal lands

(b) Total resources plus proved reserves on all lands

NOTES:

SLT = Standard Lease Terms; tcf = trillion cubic feet.

SLT category in (b) includes 42.7 tcf of gas on non-federal lands (ARI, 2001), 26.4 tcf proved reserves, and 4.6 tcf reserve appreciation.

Restricted Access category in (b) includes 6.8 tcf reserve appreciation.

Proved reserves and reserve appreciation are estimated for individual plays from the difference between undiscovered conventional or nonconventional gas (ARI, 2001, Table 2) and total gas (NPC, 1999, Appendix J, Table 9), assuming that reserve appreciation applies to conventional plays only and equals the amount of undiscovered conventional reserves, consistent with U.S. Geological Survey data for Rockies in Table 4.1.

Gas from reserve appreciation in (b) (11.4 tcf) is divided between Restricted Access (60 percent) and SLT (40 percent), according to NPC estimates (NPC, 1999).

Figure 3.2—Effect of Including Non-Federal Lands and All Gas Resources on Access Levels in the Greater Green River Basin

estimates of the potential impact of access restrictions on gas supply and prices do not depend on this distinction. These effects depend only on the fraction of total resource that is subject to restrictions. Thus, the calculations presented by the National Petroleum Council (1999) illustrating the effect on supply and prices of increased and decreased access restrictions remain valid despite the fact that they are based on the technically recoverable resource base.

Beyond the resource base upon which the access restrictions are applied, there are other aspects in the existing access restriction analyses that could be improved upon. These include assumptions about restriction exemptions, technology, infrastructure, and restriction workarounds. These assumptions, along with issues with the resource basis discussed above, lead us to make the following recommendations.

Recommendations

- *Consider Only the Restricted Portion of the Viable Resource*
  Access restrictions effectively apply only to resources that could be extracted were the restrictions not in place. Thus, it is important to consider not only the technically recoverable resource, but the economically recoverable and infrastructure-supported resource when discussing legal access restrictions.

- *Evaluate Access Restrictions in the Context of All Resources*
  Access restrictions should be evaluated in the context of all resources available to industry. Both studies exclude proved reserves in their published estimates of the accessible resource base; these reserves are substantial (14 to 21 percent of the gas in the Rockies; Table 2.2) and not subject to access restrictions. In addition, the Greater Green River Basin study includes only federal lands in the resource base, despite the fact that over 25 percent of the gas in the basin lies under non-federal lands, which are not subject to access restrictions.

- *Account for Stipulation Exemptions*
  Federal land management agencies determine which stipulations apply to a given lease on a case-by-case basis and typically record exemption requests and grants. The proportion of stipulations that are exempted can therefore be considered when estimating their impact on access to gas and oil resources in the Rockies. The Greater Green River Basin study finds that exemptions for three important types of stipulations (big game, raptors, and sage grouse) are granted in 20 to 30 percent of cases. These exemptions are included in the sensitivity case (a re-analysis with more liberal assumptions). Continuing efforts should include exemptions in the primary analysis.

- *Account for Directional Drilling and Other Low Environmental Impact Technologies*
  Technologies that allow access to resources without violating the stipulations will reduce the amount of restricted resource. One important example is directional drilling. Despite the common use of directional drilling to reach horizontal distances of 18,000 feet in Alaska (National Petroleum Council, 1999), the existing analytical approach considers a horizontal reach of less than 1500 feet and only in the sensitivity case analysis. In principle, directional drilling can be used to recover resources in regions where access is available on the scale of a few miles. Based on the distribution of accessible and restricted lands in the Greater Green River Basin (Advanced Resources International, 2001), a substantial fraction of nominally inaccessible gas may be recoverable with directional drilling.

In addition, the gas and oil industry has made large strides in developing alternative technologies to reduce environmental impacts (e.g., see Department of Energy Office of Fossil Energy, 1999). Studies of access restrictions could investigate the potential for alternative technologies to increase access to gas and oil resources within the constraints that lease stipulations are designed to maintain.

- *Include Access Restrictions on Pipeline and Road Development*
  Gas and oil development requires roads to transport equipment and personnel to the drill site and to transport the extracted resource away. The development of this infrastructure requires securing right-of-ways and construction permits, activities that may also be subject to various access restrictions. Such restrictions may preclude development even in areas where drilling is otherwise permitted.

- *Account for "Workarounds" for Existing Restrictions*
  In some cases, industry can use workarounds to access resources nominally subject to access restrictions. However, they would be likely to result in additional costs. One important option is multiple-season drilling. Lease stipulations mandating time limitations generally apply to drilling only. Once drilling is completed, production operations can typically proceed unimpeded by stipulations. This provides motivation to pursue drilling activities, even if they are delayed and incur increased costs. One common option is to drill during unrestricted time periods over multiple seasons to complete a well. Thus, even when lease stipulations preclude single-season drilling, resources can be accessed. This approach is likely to be pursued preferentially in the most resource-rich areas, and hence could allow access to substantial resources that would be recorded as inaccessible with the existing analytical approach.

4. Economically Recoverable Resources

In terms of influencing the amount of available resource, the economic recoverability criterion is expected to have the greatest impact. Assessing economic recoverability involves balancing the costs of exploration and development with the anticipated value of the resource to determine if its extraction is economically justified. Such analyses are typically conducted at the province or basin level, although the U.S. Geological Survey assesses nonconventional resources at the play level. Important factors that influence the economic recoverability of a given deposit include the kind of resource, type of formation, drilling depth, and the market price of the resource. While this type of analysis is routinely conducted by industry for planning purposes, economic assessments are rarely published. This section discusses approaches used to make such calculations.

Note that economic assessments traditionally evaluate the costs associated with getting the resource to the wellhead.[1] They do not include the costs associated with transporting the resource from the wellhead to the market. This issue of transportation is addressed as a matter of infrastructure in Section 5 of this report.

Methodology

Methodologies for conducting economic assessments include the Hydrocarbon Supply Model, used by the National Petroleum Council, Gas Technology Institute, and others (Vidas et al., 1993) and the economic component of the U.S. Geological Survey National Assessment (Attanasi, 1998). The modeling approaches are similar, although there are important differences in the data and assumptions used in each. For undiscovered conventional resources, finding-rate functions are used to predict the number of fields of a given size that will be found in individual increments of exploratory wells. The finding-rate functions are based on the recoverable resource estimates and are tailored to individual provinces or plays based upon historical discovery rates. The costs associated

---

[1]The wellhead is the point at which the resource exits the ground. Following historical precedent, the price for resource production is labeled as "wellhead," even though the cost is now generally measured at the lease boundary. In the context of domestic price data, the term "wellhead" is the generic term used to reference the production site or lease property.

34

with finding, developing, and producing the resource discovered in an increment of drilling are then subtracted from the expected revenues from production to estimate the net present value of the resource. If that value is positive, development and production are assumed to occur, and the resource is deemed economically recoverable.

Exploratory drilling is assumed to proceed incrementally in batches ranging from about 50 to a few hundred wells per batch. A key assumption in the models is that, for successive drilling increments, average discoveries are smaller and/or deeper, resulting in increasing costs. Resources are deemed economically recoverable for each drilling increment in which the costs—including exploration and development drilling costs, completion costs, lease equipment, operating and maintenance costs, all taxes and royalties, administrative costs, and a 12 percent return on investment—do not exceed the value of the resource discovered in that increment. Costs vary depending on whether a deposit is oil or gas, conventional or nonconventional, and on its depth, location, well production profiles, and by-products. The primary sources for costs are the annual Joint Association Surveys for drilling costs and the Energy Information Administration compilations of oil and gas lease equipment and operating costs. The approach is shown schematically in Figure 4.1.

In the U.S. Geological Survey analysis, successive drilling increments in conventional deposits were allowed to be targeted to specific locations and depth

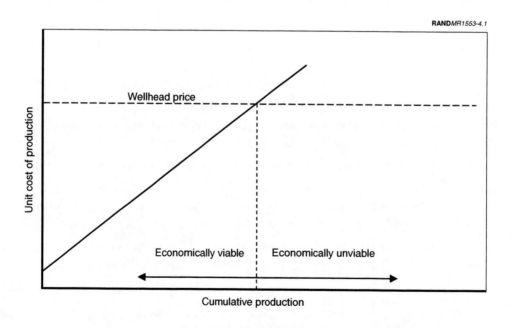

RAND*MR1553-4.1*

Figure 4.1—Economic Recoverability as a Function of Cumulative Production

intervals ("sweet spots") to maximize economic discoveries. For nonconventional deposits, however, it was assumed that industry did not have the site-specific or specialized knowledge that would allow it to selectively identify richer drilling locations. This assumption may not be valid in some localities, and drilling costs may be reduced by avoiding noncommercial wells (Attanasi, 1998).

In addition to undiscovered conventional and nonconventional gas and oil resources, a substantial fraction of the unproved resources identified in the resource assessments comes from reserve appreciation (see Section 2). Reserve appreciation presents a difficulty for cost estimation, because the modeling approach for estimating reserve appreciation is statistical rather than geological. The U.S. Geological Survey concluded that costs associated with reserve appreciation could not be estimated, because past and projected future growth could not be related to specific development efforts. As an alternative, in order to approximate potential reserves expected to be available for production in the next two decades, the U.S. Geological Survey includes reserve appreciation projected for 1994 through 2015 to the economically recoverable resource (in contrast, the technically recoverable assessment includes reserve appreciation through 2071).

The Hydrocarbon Supply Model, on the other hand, assesses the economic recoverability of resources available through reserve appreciation by estimating the number of wells required to explore and develop those resources. These wells are then subjected to the same type of economic assessment as described above for undiscovered conventional resources, with the exception that many of the exploration costs associated with new field development are excluded. This distinction is made because reserve appreciation applies to reserves from existing fields, for which many of the exploration costs have already been incurred. The economically recoverable price for resources from reserve appreciation is thus lower than for undiscovered conventional resources.

Results

Organizations differ in the way they apply the economic assessment methodology. The results are difficult to compare, because most organizations do not publish explicit price-supply relationships that would provide an estimate of the economically recoverable resource at a given wellhead price. The U.S. Geological Survey presents the most explicit illustration by publishing the technically recoverable and economically recoverable estimates separately, allowing a direct comparison between the two. The following discussion therefore focuses on the results of the U.S. Geological Survey economic assessment. It is important to note, however, that as with its technically recoverable estimate, the U.S. Geological Survey's economic assessment accounts for current technology only. As a result, its economic assessment is generally considered to be more conservative than those used by industry.

Economic assessments were carried out for two price levels, one assuming $18 per barrel (bbl) of oil and $2 per thousand cubic feet (mcf) of gas, and the second at $30/bbl oil and $3.34/mcf gas.[2] The results are summarized in Table 4.1. Based on the mean values from the 1995 U.S. Geological Survey National Assessment, the results indicate that, nationwide, 53 percent of the recoverable oil and 38 percent of the recoverable gas is economically extractable at $18/bbl and $2/mcf. At $30/bbl and $3.34/mcf, the economically recoverable fractions of oil and gas rise to 61 percent and 46 percent, respectively. Note that these figures include proved reserves, which, by definition, are economically recoverable.

These results illustrate the substantial role that economic considerations may play in determining the available gas and oil resource. Even at the higher prices analyzed, 40 percent of the recoverable oil and 55 percent of the recoverable gas are effectively inaccessible simply because they cannot be extracted profitably. The economics are even more restrictive for nonconventional resources. Although nonconventional gas represents the largest single category of recoverable gas nationwide, only 21 percent of nonconventional gas is economical to develop (Table 4.1). Thus, while nonconventional gas resources are substantial, they are generally more expensive than conventional resources to extract, and any discussion of the availability of gas resources must clearly distinguish between conventional and nonconventional resources. Nonconventional oil is similarly more expensive to extract, although it represents a much smaller fraction of the total oil resources.

This distinction between conventional and nonconventional resources is particularly relevant in the Intermountain West region. Within the Rocky Mountains and Northern Great Plains (U.S. Geological Survey Region 4), 81 percent of the unproved recoverable gas resource is in nonconventional deposits. Of this, only 8 percent is estimated to be economically recoverable at $3.34/mcf, and only 4 percent is recoverable at $2/mcf (Table 4.1). Areas within the Intermountain West region show even greater fractions of nonconventional

---

[2]Note that prices reflect 1997 dollars. Gas price is assumed to be two-thirds the oil price on an equivalent energy basis. At an energy equivalent of 1 bbl oil = 6 mcf gas, the gas price per mcf = 0.111 x oil price per barrel. This relationship between gas and oil prices corresponds roughly to the historical average (Attanasi, 1998).

Table 4.1

Economically Recoverable Oil and Gas in the United States (USGS)

	Technically Recoverable[a]			$18/bbl or $2.00/mcf			$30/bbl or $3.34/mcf		
	Oil (BB)	Gas[b] (tcf)	NGL (BB)	Oil (BB)	Gas[b] (tcf)	NGL (BB)	Oil (BB)	Gas[b] (tcf)	NGL (BB)
United States									
Conventional	30.3	258.7	7.2	9.2	77.5	3.0	17.4	121.8	4.6
Nonconventional[c]	2.1	358.0	2.1	0.14	35.9	0.15	1.1	74.4	0.20
Reserve appreciation	60.0	322.0	13.4	30.3	160.0	-	30.3	160.0	-
Proved reserves	20.2	135.1	6.6	20.2	135.1	6.6	20.2	135.1	6.6
Total	112.6	1073.8	29.3	59.84 (53%)	408.5 (38%)	9.8 (33%)	69.0 (61%)	491.3 (46%)	11.4 (39%)
Rockies & N. Great Plains (Region 4)									
Conventional	4.63	21.91	1.45	1.73	11.4	1.1	2.98	15.0	1.4
Nonconventional[c]	0.45	173.0	1.73	0	6.9	0	0.04	14.0	0.05
Reserve appreciation	6.8	19.2	0.9	2.8	10.5	-	2.8	10.5	-
Total unproved	11.88	214.11	4.08	4.53 (38%)	28.8 (13%)	1.1 (27%)	5.82 (49%)	39.5 (18%)	1.45 (36%)
SW Wyoming (Prov. 37)									
Conventional	0.17	1.58	0.02	0.02	0.16	0	0.08	0.59	0.01
Nonconventional[c]	0	123.1	1.73	0	1.09	0	0	5.11	0.05
Total undiscovered	0.17	124.68	1.75	0.02 (12%)	1.25 (1%)	0 (0%)	0.08 (47%)	5.7 (5%)	0.06 (3%)

SOURCES: U.S. Geological Survey National Oil and Gas Resource Assessment Team (1995); Attanasi (1998); Root et al. (1997).

[a]Technically recoverable category shows the mean values reported in the U.S. Geological Survey 1995 National Assessment and includes onshore and state offshore resources, but excludes federal offshore areas.

[b]Gas values include associated and non-associated gas.

[c]Nonconventional gas includes coalbed methane.

NOTE: bbl = barrel, BB = billion barrels, mcf = thousand cubic feet, tcf = trillion cubic feet, NGL = natural gas liquids.

38

resources. In southwestern Wyoming (U.S. Geological Survey Province 037), for example, over 99 percent of the undiscovered conventional gas resource is in nonconventional deposits. Only 4 percent of this nonconventional resource is economically recoverable at $3.34/mcf. Thus, while gas resources in the Intermountain West are abundant, they are primarily contained within nonconventional deposits and are therefore largely inaccessible for economic reasons.

The data and forecasting assumptions used in Table 4.1 are current as of about 1994. It is important to note that technological improvements and changing economic conditions will alter these estimates over time. The use of more current recoverable resource estimates and cost assumptions will undoubtedly alter the results, particularly regarding the costs of developing nonconventional resources. Technology in this area is progressing rapidly, and the economically recoverable fractions are likely to be higher today than reported in Table 4.1. Nonetheless, nonconventional resource development remains substantially more expensive than conventional resource development.

Recommendations

Existing approaches were developed for the purposes of making assessments on a national scale; they do not necessarily contain the resolution needed to properly address many of the specific issues related to the individual basins in the Rockies. In updating evaluations of the economically recoverable resource in the Intermountain West, improvements can be made to the standard economic models to help tailor our economic evaluation to account for some of the characteristics of the region and to improve the accuracy of economic modeling of resource development.

- *Use Data That Reflect the Region of Interest*
  Significant characteristics of gas and oil development in the Rockies can vary depending on the size and nature of the basin. Understanding the costs therefore requires cost data tailored to each basin. However, cost data available through the Joint Association Survey and the Energy Information Administration are generally presented by state. Further, the Hydrocarbon Supply Model (used by the National Petroleum Council) aggregates multiple state data to generate average costs for larger regions. The National Petroleum Council's economic considerations regarding the resources in the Rockies are thus based on regional average costs that may not reflect the actual costs of extracting gas and oil from many of the basins in the region.

To accurately estimate costs for Rocky Mountain gas and oil, the cost data should be based on data from the specific areas being considered.

- *Account for the Abundance of Nonconventional Gas in the Rockies*
  A very high fraction of undiscovered gas in the Rockies is contained in nonconventional formations. This distinction is expected to impact costs for well completion, lease equipment, and operating costs. However, the National Petroleum Council cost estimates are regional averages based upon existing wells, the majority of which are in conventional deposits. These estimates may underestimate the real costs of developing Rocky Mountain gas and oil. To account for the higher costs associated with nonconventional formations, the Hydrocarbon Supply Model includes an additional stimulation cost for low-permeability wells. In addition, the Joint Association Survey includes estimates for coalbed methane drilling costs. However, these estimates again are aggregate values, in the latter case representing average costs for the entire nation. Costs should reflect the real costs associated with nonconventional resource extraction in the Rockies basins, including the distinctions between tight sandstone and coalbed methane.

- *Use Local Drilling Success Rates*
  The high fraction of nonconventional deposits may also influence drilling success rates. The drilling success rate is the fraction of drilled wells that are productive and influences the total number of wells that must be drilled. As with the costs, the drilling success rates used in existing assessments reflect regionwide averages of existing wells and are therefore strongly biased toward conventional deposits. A meaningful economic recoverability assessment should be based upon the best estimates for drilling success rates for the specific basins and specific types of deposits being considered.

- *Address Other Costs Specific to the Rockies*
  Other aspects of the Rocky Mountain Region may also influence resource extraction costs compared to other regions. These include the steep and rugged terrain, the sparse pipeline and road infrastructure, and above-average costs associated with environmental impact assessment and mitigation. Conversely, the shallow depth of many formations reduces costs. All of these factors should be considered in assessing the costs of gas and oil produced in the Rocky Mountains.

- *Determine the Economically Recoverable Resource Explicitly*
  Accurate cost and performance data should be compiled and used to construct cost algorithms that include the individual cost components discussed above. These cost algorithms can then be combined with existing estimates of technically recoverable resources to generate price-supply

relationships for individual plays. Such relationships, analogous to those derived in the U.S. Geological Survey economic assessment (Attanasi, 1998), will provide an estimate of the fraction of the technically recoverable resource that can be extracted profitably at any specified wellhead price.

5. Infrastructure

A critical aspect of gas and oil production is the infrastructure necessary to provide access to the drilling site and to move the extracted resource from the well to the market. Key considerations are proximity of existing pipeline infrastructure, pressure under which the resource is produced, presence of contaminants, and proximity of consumer markets. The costs of new pipeline, compression, processing, and transmission infrastructure impact the economics of resource development negatively. In many cases resources will still be economically recoverable with these additional costs. In other cases, however, these costs can be prohibitively high. The availability of infrastructure thus represents an additional criterion for defining the economically recoverable resource.

Typically, resource assessments do not consider the infrastructure requirements associated with resource development. Including this element is important for assessing the viable resource in the Rocky Mountains, however. Compared to other parts of the country, the Rocky Mountain Region is generally farther from consumer markets, and has considerably less road, pipeline, and processing infrastructure. In addition, because of the remoteness of the area and the rugged terrain, construction, operation, and maintenance of infrastructure is technically challenging and more costly than in many other parts of the country.

In this section, we outline an approach for including infrastructure requirements into the viable resource definition. We first provide a brief overview of infrastructure components. We then discuss how we can account for these operating costs when making a viable resource estimate. Our approach applies to the production of natural gas and oil, although our discussion focuses on nonconventional natural gas, including coalbed methane and tight sandstone.

Overview of Infrastructure Components

In this section, we discuss the infrastructure and equipment necessary to move the resource from the well to the market. We start at the equipment around the wellhead and continue with the local pipeline infrastructure (the "gathering system"), the processing facilities, and the interstate transmission pipelines. Table 5.1 provides an overview of these elements, with notes on specific aspects of production in the Rocky Mountains. Note that the distinction between

42

Table 5.1

Infrastructure Components, Cost Items, and Issues Specific to the Rocky Mountains

Infrastructure Component	Cost Items	Specific Issues
Water disposal	Costs for disposal; flowlines, re-injection well	Relatively high water production, especially coalbed methane
Compression system	Compression systems, fuel consumption	Coalbed methane produced at low pressure
Gathering system	Flowlines, gathering lines	Less economies of scale through fewer existing systems; lower flow rates per well; worse terrain conditions
Processing facility	Dehydration, CO2, N2	More contaminants in nonconventional resources; fewer economies of scale through terrain conditions
Access to transmission pipelines	Transmission charges + fuel consumption	Long distance to markets

wellhead and infrastructure cost elements can be ambiguous. Flowlines, for example, are often considered part of the wellhead costs. We include them here for completeness. They will be assigned to either wellhead or infrastructure costs when we complete our cost analysis.

Water Disposal

Produced water is the largest byproduct of resource production. While natural gas wells typically produce much lower volumes of water than oil wells, a notable exception are certain types of gas resources such as coalbed methane or Devonian/Antrim shales. In the case of coalbed methane, aquifers must be pumped out ("produced") to cause coalbeds to release methane gas. Common methods of disposal are surface disposal or, more commonly, reinjection. Over 90 percent of onshore produced water is disposed of through injection wells or for use in water-flooding for enhanced recovery (American Petroleum Institute, 2000). Reinjection is expensive, with the creation of reinjection wells being the major cost factor. In some cases, produced water and other wastes may need to be removed from the site, potentially requiring additional pipeline capacity.

Compression

Some gas is produced at much lower pressures than required for transmission. The transmission pipelines require some 800 to 1100 psi (pounds per square inch). Tight sandstone is usually produced at around 700 to 800 psi. Coalbed methane, however, is produced at much lower pressures, usually around 10 psi. Therefore, the gas needs to be compressed before feeding it into the transmission pipeline. The maximum compression ratio (outlet pressure/inlet pressure) of typical compressors is four. Consequently, coalbed methane gas may need to be compressed in four stages (from 10 to 40 psi, from 40 to 160 psi, etc.) to reach the pressure of the transmission pipelines.

Gathering System

The gathering system is a network of pipelines that moves gas from individual wells to compressor stations, processing points, or transmission pipelines. The extent of the gathering system depends on well spacing, production volumes, and the location of transmission pipelines. A gathering system may consist of hundreds of miles of pipelines connected to a hundred or more wells in a single field. The gathering system consists of flowlines and gathering lines.

Flowlines are tied to individual wells or equipment (located at or near the well site), which move wellhead fluids or gas to the first point of accumulation. In small oil or gas fields, flowlines typically serve one wellhead. In multiple well fields, producers commonly lay flowlines from individual wells to a central facility to perform future production processes. The flowlines are usually no longer than two miles.

When natural gas is produced with crude oil, both share the initial surface flowline from the wellhead to the gas separator. From the separator onward, the natural gas is transported in its own pipeline system, except in some special installations where two-phase pipelines are used (Berger and Anderson, 1992).

Gathering lines are the next segment of the gathering system. There are two types of gathering lines. First, there are gathering lines that connect flowlines with the central collection point or a processing facility. These gathering lines are tied to the flowlines through an intermediary manifold. If separation, treating, heating, dehydrating, compression, pumping, or other processing has not occurred along the flowline before the fluid or gas is gathered, then the gathering lines will transport the fluids or gasses through a processing point such as a central facility. Second, there are gathering lines that connect the processing facility or central collection point to the interstate transmission pipeline.

The costs of gathering systems in the Rockies may be higher than in other areas. Producers cannot as easily combine gathering systems. Often, companies integrate their systems to save costs. However, there is not as much existing pipeline infrastructure in the Rockies as in other, more mature areas. Therefore, economies of scale in gathering systems are relatively low in new areas like the Rocky Mountains. The terrain conditions tend to magnify this effect by isolating production areas. Gathering systems are also more extensive as nonconventional resources tend to have more wells and lower flow rates than otherwise identical conventional wells.

Processing

Depending on the gas characteristics, gas processing may involve dehydration, removal of natural gas liquids, or removal of impurities. Dehydration is often necessary because of requirements for transmission and production at low outside temperatures. Normally, owners of transmission pipelines limit the amount of water that can be contained in gas to seven pounds per million cubic feet. Low outside temperature may lead the hydrates to freeze, clogging up the gathering system. Dehydration often occurs at the wellhead. It is a relatively simple and inexpensive process.

Other contaminants may have to be removed as well. Many Rocky Mountain reservoirs contain carbon dioxide and/or hydrogen sulfide. These contaminants must be removed unless present in very small concentrations. Coalbed methane in particular tends to contain high amounts of carbon dioxide. Many deep reservoirs contain both carbon dioxide and hydrogen sulfide. Inert gases (primarily nitrogen) must be removed in an expensive process or blended with other gas.

Transmission Pipelines

The gathering system connects to the interstate transmission pipelines. The gathering system connects to the interstate transmission pipelines, which move gas from producing regions to consuming regions. The diameter of these pipelines ranges from 20 to 42 inches and the gas pressure typically ranges from 800 to 1100 psi. Many major interstate pipelines are "looped"—there are two or more lines running parallel to each other in the same right of way. Compressor stations are located approximately every 50 to 60 miles along each pipeline to boost the pressure that is lost through the friction of the natural gas moving through the steel pipe.

The Rocky Mountain area suffers from a lack of receipt or pipeline exit capacity at expanding production areas. Rising production levels in Wyoming's Powder River basin, as well as in several other Rocky Mountain production areas, are placing pressure on local pipeline systems and regional transmission pipelines to expand their capabilities to move more gas to nearby and distant markets (Energy Information Administration, 2000).

Roads

Because of the remoteness of many areas of potential production, the network of roads is much less developed in the Rocky Mountains than it is in other areas.[1] Therefore, companies in the Rockies will have to construct new road infrastructure, resulting in higher development costs.

Producing companies will have to create road access to the well, using existing road infrastructure and/or building new infrastructure. The length of new road infrastructure is dependent upon the well site location in relation to existing roads or highways.

Including Infrastructure in Resource Assessments

The infrastructure criterion is introduced to identify those resources for which sufficient infrastructure is available or can be constructed within economic constraints. Making this judgment entails determining infrastructure requirements, estimating costs for augmenting existing infrastructure to meet those requirements, and evaluating whether the expected value of the resources produced would be sufficient to justify the infrastructure costs.

Infrastructure Requirements

In general, infrastructure requirements can be estimated from resource production characteristics, location, and terrain. A simple framework for estimating infrastructure requirements is illustrated in Figure 5.1. In assessing infrastructure needs, we distinguish road, gathering/processing system, and transmission infrastructure and costs.

Our proposed approach to estimating the infrastructure costs begins with estimating the number and locations of wells needed to extract the total gas and oil resource from individual plays and basins in the Intermountain West. The number of wells needed to extract resources from a given play depends on a

---

[1]Population per square mile in Wyoming = 5, Montana = 6, and Colorado = 32, versus 70 for all of the United States (Demographia, 2002).

46

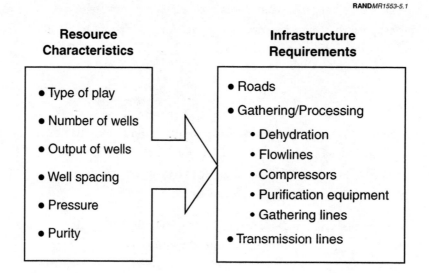

Figure 5.1—Framework for Estimating Infrastructure Requirements

number of factors, including well spacings, recovery per well, and formation type and pressure, and can be highly variable. Well spacing is often legally defined by the producing state. For example, the coalbed methane well spacing in the Raton Basin dictated by the Colorado Oil and Gas Conservation Commission is typically one well every 40 acres, though this spacing can range from as little as 2.5 acres to as much as 160 acres (East of Huajatolla Citizens Alliance, 2002). Legally governed well spacings are typically established based upon divisions of governmental sections and are often not technically meaningful. Horizontal drilling can also reduce the number of necessary wells drilled. Estimating the number of wells needed for a given play thus requires consideration of geological, regulatory, and technological factors.

The infrastructure requirements associated with the required wells can be estimated on a per-well basis from a parameterization of the expected distribution of wells, their flow rates, specific equipment requirements, and distances from transmission pipelines and roads.

Road networks needed to reach each well can be estimated from the proximity to existing road networks. The U.S. Department of Transportation maintains databases of road locations and types. From these the additional roads that would need to be built to reach each well can be estimated. The access road built to the site is generally a simple procedure but is complicated by rough terrain and icing issues in colder climates. Unstable soil may necessitate board roads to be built and footings made of gravel.

The diameter and length requirements of flowline systems are determined by the expected capacity output of the well and the proximity to the gas processing plant. Gas processing plants can service multiple wells, and the number of plants necessary to service a given play will depend on the size and type of play in addition to the quality of the natural gas found. The impurities expected will also control the type of processing plant needed and associated costs. Some currently existing gas processing plants may be able to handle additional throughput. In this case, longer flowlines may be used to reach these facilities in lieu of building new processing capacity.

The size and cost of a gathering system vary considerably and depend on the size of the resource, the proximity of the proposed wells to existing transmission lines, and terrain features. Requirements for the Intermountain West can be estimated from existing systems and industry data and summed to cover projected play resources. As development progresses, some of the load increase could be supported by excess capacity on the existing infrastructure. Subsequent growth would necessitate the construction of new systems. An important consideration for modeling new gathering systems is estimating the capacity of future increments. For example, for a given development increment, too large a system would never be economically feasible, while too small a system would reach maximum capacity quickly. A reasonable approach would be to match the proposed gathering system capacity with the requirements generated in the drilling increments used in the economic assessments.

The final infrastructure component is the transmission line. The extent to which existing lines can support continued development in the Intermountain West depends on several factors, including the available capacity on the line, the nearest point of possible interconnect, transmission charges, and impacts on constrained interfaces (bottlenecks). These issues can be assessed for different locations in a region and the costs associated added to the cost of bringing the resource to market.

Other Considerations

Rugged and steep terrain can make infrastructure development difficult and expensive. Road and pipelines built over excessively rough terrain incur additional costs and can be prohibitively difficult to construct. High-slope areas are defined by grades of greater than 30 percent, consistent with lease stipulations in the western United States. They are considered difficult areas for exploring, drilling, and laying pipelines. These slopes can be calculated from the USGS-supplied digital elevation model (U.S. Geological Survey, 2002) and the areas that meet some terrain criteria can be evaluated against current technologies.

The Viable Resource

For an increment of wells with a given set of production characteristics, the distance to a transmission pipeline can be used to estimate the infrastructure required to process and transport the resource to the transmission line. The location and capacity of existing infrastructure can be used to estimate the additional infrastructure that is needed to be built to accomplish this.

In conjunction with wellhead costs, the costs of the required infrastructure for an increment of wells can be then compared to various assumed market prices to determine if the costs of building the infrastructure would likely render development to be unprofitable. Such an approach can be used to identify resources that are located close enough to transmission pipelines to be included in the viable resource base.

A preliminary estimate, based primarily on input from industry executives and published costs, indicates that if all new infrastructure is required, the additional costs for natural gas development in some areas of the Rockies could be more than 50 percent of the wellhead costs. In many cases, resources may still be economically recoverable when the additional infrastructure costs are accounted for. In other cases, however, the infrastructure requirements may prevent an otherwise attractive development from proceeding.

Using GIS analysis, the productive areas of individual plays can be overlaid with a data layer of existing transmission lines to assess the percentage of resources that are within a given distance of a major transmission line. Productive areas will be assigned to separate categories according to expected production characteristics to distinguish areas with different infrastructure requirements. Data on the location of existing gathering lines and gas processing stations will determine well locations within a given distance of current infrastructure. Furthermore, terrain databases can limit the areas of erecting infrastructure even further.

The results of this analysis would be the amount of resource that, based on distance from transmission lines, could be economically developed.

6. Environmental Considerations

Impact on the environment is an important consideration in estimating the viable resource. As discussed above, recent studies have examined the impact of environmental constraints on drilling on federal lands by assessing access constraints imposed through lease stipulations (e.g., National Petroleum Council, 1999; Advanced Resources International, 2001). The scope of these analyses is limited to evaluating the effects of lease stipulations as currently written. They also do not consider the environmental impacts of development on non-federal land. This approach thus provides a useful summary of the current regulations that parties interested in resource development on federal lands may encounter today. However, federal agency regulations and land use classifications are not static: in some cases exemptions for lease restrictions may be granted[1] and in other cases restrictions can be imposed upon nominally unrestricted land. In addition, access restrictions are subject to differing and potentially inconsistent objectives, interpretations, and capabilities of several different federal land management agencies. As biological and geological surveys are updated, as technologies improve, and as the political climate changes, land may be reclassified and lease stipulations may be changed.

An alternative approach to incorporating environmental impacts into estimates of the viable resource is to consider physical, chemical, biological, and ecological properties of the land irrespective of lease stipulations. Changes in these measures would be used represent potential impacts of oil and gas extraction. The goal of this analysis is to estimate the amount of oil and gas resource under lands of varying levels of sensitivity to changing conditions.

In this section, we discuss some potential extraction-related environmental stresses that might be included in building such a classification. In this discussion of potential impacts, there is no intention to estimate magnitude of impact, or to make judgments as to whether impacts will occur with or without mitigation efforts. This is simply a discussion, based on the literature, of the

---

[1]Complete data on the incidence of stipulation exemptions are not available, but some local data are available. For example, during the 2001–2002 winter (data available through February 21, 2002), the Pinedale, Wyoming BLM Field Office granted 64 percent of requests for exemptions and denied 21 percent (10 percent were withdrawn and 3 percent each were not required or are pending) (Keith Andrews, Pinedale, Wyoming BLM Field Office, personal communication).

potential unmitigated impacts of oil and gas extraction based on current
technological practices (U.S. Environmental Protection Agency, 2000).

Overview of Environmental Impacts

Oil and gas extraction involves exploration, well development, production,
maintenance, and site abandonment. It also involves transport and waste
disposal. In each of these stages, use of various equipment and methods, as well
as introduction and release of various materials, create several demands on the
environment, as summarized in Table 6.1. In detail, impacts of oil and gas
extraction projects may vary considerably as a consequence of site-specific
conditions and the specific exploration and development processes used. In
general, however, the scope of environmental impacts can be considered along a
number of dimensions:

- environmental features and processes
- duration of impact
- direct versus indirect impacts
- magnitude of impact
- available mitigation to reduce impact.

The first consideration is the type of environmental feature or process being
impacted. Perhaps the most obvious impacts resulting from oil and gas
extraction are ecological (i.e., related to the natural functions and relationships
among organisms, air, water, and land). These begin with the construction of the
drilling apparatus, service roads and pipelines. In addition, solid and hazardous
waste and large volumes of wastewater are generated during construction,
operation, and abandonment of the project, with potential implications for
regional air, water, and soil quality. Finally, though major accidents are
relatively rare, the effects of spills and blowouts are potentially serious.
Together, these disruptions have the potential for adversely affecting complex
ecosystems. Ecosystems in the Rocky Mountain area include a rich array of plant
and animal species, yet face increasing population growth and continued stresses
associated with ongoing anthropogenic processes including agriculture and
grazing, urbanization, and mineral extraction (Mac et al., 1998).

Environmental impacts may also extend beyond the ecological to include impacts
on resources of cultural, historical, and societal value. Anthropological and
paleontological remains and historic buildings and districts often warrant
protection from disturbance as environmental resources. One of the more

51

Table 6.1

Potential Environmental Impacts from Oil and Gas Extraction

	Land Use, Aesthetics, and Ecological Resources	Air Quality	Water Quality	Soil Quality
Exploration and Development	Noise, vibration, devegetation, roads, traffic, habitat fragmentation, litter, construction, heavy equipment, percolation and disposal pits, growth inducement, potentially irreversible land use change, ecosystem disruption	Emission of fugitive natural gas, VOCs and PAHs, combustion emissions, hydrogen sulfide, sulfur dioxide, greenhouse gases	Surface and/or downhole release of formation and injected contaminants, alkalis, brines, acids, diesel oil, and crankcase oils	Compaction by heavy equipment and solid waste; contamination with brine, acids, chemical additives, and formation contaminants
Production	See above	Continued release of above emissions; fugitive BTEX (benzene, toluene, ethylbenzene, and xylene) from natural gas conditioning	Surface and/or downhole release of produced water possibly containing heavy metals, radionuclides, dissolved solids, oxygen-demanding organic compounds, additives, high levels of salts, untreatable emulsions, and formation fluids	Physical and toxicological impact of heavy machinery and solid waste (sometimes contaminated), spent catalysts, separator sludge, tank bottoms, used filters, and sanitary wastes

Table 6.1 (continued)

	Land Use, Aesthetics, and Ecological Resources	Air Quality	Water Quality	Soil Quality
Maintenance	See above	Emission or production of volatile cleaning agents, paints, other VOCs, hydrochloric acid, fugitive natural gas, PAHs, particulate matter, sulfur compounds, carbon dioxide, and carbon monoxide	Surface and/or downhole release of completion fluid, wastewater containing well-cleaning solvents (detergents and degreasers), paint, stimulation agents, escaping oil, and brine	Physical and toxicological impact of pipe scale, paint wastes, paraffin, cement, sand, sorbents, and contaminated soils
Accidents and Abandoned Wells	Landscaping containment structures; fires; and introduction of hazardous waste into habitat	Emission of fugitive natural gas, VOCs, PAHs, particulates, sulfur compounds, carbon dioxide, carbon monoxide	Accidental release of regulated toxic contaminants, oil, and brine	Contamination with escaped formation and injected materials, impact from containment structures and burning

SOURCE: Modified from Sittig (1978) and U.S. Environmental Protection Agency (1987).
NOTE: VOC = volatile organic compound; PAH = polyaromatic hydrocarbon.

difficult impacts to consider, but often one with the greatest public interest, particularly in scenic areas such as the Rocky Mountains, is the aesthetic impact to landscapes; introduction of machinery, development of roads, and denuding vegetated landscapes to support extraction activities clearly have potential aesthetic implications.

A second important consideration is the duration of impact. Impacts can be short term, such as the temporary displacement of plant and animal species, or long term, such as the contamination of groundwater by fluid migration in abandoned wells. Short-term activities may have short- or long-term impacts. For example, the behavior of many animal species often changes under the influence of noise (Ercelawn, 1999); disrupting certain activities, such as foraging and breeding, of threatened and endangered animals may have important consequences for species survival. In the case of oil and gas extraction, it is important to note that the long-term use of the land does not cease with site abandonment. Although abandonment often includes restoration of the site, the site may continue to be used for waste disposal, and activities such as road and pipeline development create permanent structures that may induce further development of the site and adjacent lands over time. Recent studies are beginning to document various ecological impacts associated with road development in the United States (Ercelawn, 1999).[2]

Third, impacts can be direct and/or indirect. Clearing a site of its vegetation, for example, directly impacts the land cover on the site, altering both its appearance and its function. Clearing also may create fugitive dust or increase siltation and runoff into surrounding streams. Less obvious indirect effects may include habitat fragmentation and blockage of migration corridors that, over time, affects the genetic exchange occurring in sensitive species of native plants and animals. Cultural resources can also be affected indirectly. For example, a single historic structure may not, in and of itself be valued, but its relation to others within a historic town establishes an area worthy of preservation. Some projects might lead to local impacts such as groundwater contamination, regional effects such as watershed pollution, and global effects such as natural gas emissions that contribute to climate change. An example of the latter is fugitive natural gas emitted during oil and gas extraction. In fact, methane from gas drilling and transmission accounted for 16 percent of methane from global anthropogenic sources in the 1980s; although methane accounts for only about 4 percent of

---

[2]The relationship between road development and subsequent forest clearing, for example, is a dominant theme in tropical deforestation literature, although deforestation processes in developing countries are understood to be a function of a number of additional factors less apparent in the United States.

anthropogenic greenhouse gases, it is more than 20 times more effective in absorbing infrared radiation than carbon dioxide (Houghton, 1990).

An additional consideration is the magnitude of impact. Thresholds of significance—specific levels of impact above which relevant regulations or overseeing authorities consider values to be severe enough to cause concern—are useful in measuring magnitude. Stricter standards (i.e., lower thresholds) generally apply to resources of higher value and greater sensitivity. From a biological perspective, for example, the value of critical habitat for endangered species exceeds that of landscaped areas that support more common species. Development in a critical habitat area could potentially and significantly impact sensitive native plants and animals, while development of a fallowed agricultural field might not. Existing environmental regulations aim to protect environmental resources, particularly those that are most sensitive.[3] The oil and gas extraction industry is regulated, with compliance intended to minimize impacts of extraction processes. State and federal agencies monitor the compliance of facilities and sites with various statutes, regulations, and enforcement practices. Inconsistency with applicable environmental regulations may constitute a significant impact.[4]

Lastly, impacts can be considered avoidable or unavoidable. Potentially significant impacts often can be avoided through various project redesign options, or mitigated to less-than-significant levels by following appropriate best management practices. For example, reduced footprint protocols and directional drilling near sensitive areas can reduce impacts of extraction operations. Smaller boreholes can reduce the amount of solid waste and drilling muds that must be disposed of. Off-season construction and operation can reduce disruption of nesting and breeding activities of sensitive species. Water treatment and runoff control techniques can reduce impacts of waste on surface and groundwater. Sometimes direct loss of habitat due to a project can be mitigated by payment of in-lieu fees to support a regional habitat conservation plan. In areas where emissions trading occurs, emissions credits for air and water pollution can be

---

[3]Relevant federal laws include the Clean Water Act and its subsection, the Oil Pollution Prevention Regulation; the Safe Drinking Water Act; the Solid Waste Disposal Act, as amended by the Resource Conservation and Recovery Act; the Comprehensive Environmental Response, Compensation, and Liability Act (or Superfund); the Superfund Amendments and Reauthorization Act; the Clean Air Act; the Federal Insecticide, Fungicide, and Rodenticide Act; the Toxic Substances Control Act; the Endangered Species Act; and the National Environmental Policy Act.

[4]The extent that facilities violate laws or statutes varies considerably by facility and region. An EPA database that tracked national noncompliance history in the oil and gas sector from 1992 to 1997 found that the ratio of enforcement actions to inspections ranged from 0 to 0.17, with a value of about 0.04 for the Rocky Mountain Region (U.S. Environmental Protection Agency, 2000). Intervals between inspections were found to range from one year in the New York region to nearly six years in the Rocky Mountains.

bought and sold in a collective agreement between firms to achieve air and water quality goals. Thus, mitigation options, if available, can serve to reduce environmental impacts, but often at additional cost to the developer. Certain mitigation options may be cost-prohibitive. Unavoidable impacts, simply, are those that cannot be mitigated to less-than-significant levels under the project. Before a project is approved, the National Environmental Policy Act (NEPA) requires that impacts on the environment be disclosed in a public review. Further, certain states require that projects with significant negative environmental impacts produce a statement of overriding considerations that explains why the social benefit of a project exceeds its environmental harm.

Exploration and Development

The exploration process involves non-invasive exploration techniques as well as exploratory drilling. There are several types of non-invasive methods including satellite remote sensing and surface geophysical techniques, such as generation of seismic waves by detonating explosives in subsurface holes or dropping heavy weights on hard surfaces (land vibreosis). Remote sensing and surface geophysical techniques are considered relatively non-invasive means of ascertaining subsurface rock properties. Potential negative environmental impacts of seismic techniques are primarily associated with noise and vibration, which can disturb people in nearby populated areas and activities of animals (e.g., foraging and breeding, which may be of greater concern to areas that support sensitive animal species during certain times of the year). Exploratory drilling is always required during exploration to confirm the presence of oil and gas, but generally requires a smaller borehole and generates less waste than during production well development.

During drilling, drilling fluid (mud) is pumped down the borehole and back to the surface to lubricate the bit, remove rock fragments, and maintain pressure that keeps the wellbore intact and free of oil, gas, and water. Spent muds contain a number of additives used to enhance the muds' performance. They also may contain chemical contaminants associated with the underlying formation, including heavy metals and hydrogen sulfide. These additives and chemicals—which can be toxic to nearby plants and animals and harmful to equipment—must be removed from the fluids and disposed of, often by costly methods.

Rock fragments (cuttings) brought to the surface in the drilling fluid pose problems because of their high volume and because they are coated with contaminated mud. In addition, solid wastes such as cement and metal may be left over from the casing process. It has been estimated that between 0.2 and 2.0

barrels (8.4 and 84.0 gallons) of drilling waste are produced for each vertical foot drilled (U.S. Environmental Protection Agency, 1987). The level of difficulty in solid waste disposal differs greatly among sites, and disposal is regulated by various federal laws, with compliance aimed at minimizing impacts to air, water, and land.

Water-based muds and associated cuttings can be discharged to the surrounding areas including surface waters in accordance with applicable environmental regulations. Oil-based muds and associated cuttings must be conditioned and reused, or disposed of either by treatment, incineration, or on-site burial, or must be transported for off-site disposal. Improper disposal and subsequent leaching of hazardous waste—including oil, brine, and acidic solutions—into surface and groundwater supplies can cause a number of harmful effects on surrounding vegetation, aquatic organisms, and higher animals.

The drilling operation results in air emissions typical of combustion, such as exhaust from diesel engines and turbines that power the drilling equipment. Hydrogen sulfide is also commonly a by-product of the drilling process, which, as described above, is toxic to various plants and animals. Incineration of hydrogen sulfide can produce sulfur dioxide, an air pollutant found in acid rain.

Exploration and drilling activities also result in the accumulation of domestic and sanitary wastes from sinks, showers, laundry, food preparation areas, and toilets. Various federal laws regulate fluids and waste management, with compliance aimed at minimizing impacts to surface and groundwater. Wildlife are generally restricted from the area and exposure to hazardous materials.

In most cases, air, water, and wildlife impacts can be mitigated to less-than-significant levels, but the cost of this mitigation needs to be considered in the resource estimate.

Production

If tests indicate that sufficient hydrocarbons are present, a well is completed and production begins. In cases of poor flow, hydraulic fracturing (introducing liquid at high pressure) and/or acidizing (pumping acid, most frequently hydrochloric acid, to dissolve soluble materials) may be employed to open pores and stimulate flow. If well stimulation is necessary, the clogging paraffin and any other acid-dissolved or pressure-fractured solid waste brought to the surface from the formation must be disposed of. Production may also involve injecting water to repressurize the reservoir, injecting gas to enhance gas cap drive, and injecting oil-miscible fluids, surfactants, and organic-digesting microbes to mobilize lighter oil or gas.

The largest-volume by-product of the production process is produced water. A 1995 survey found that 15 billion barrels are produced annually (American Petroleum Institute, 2000). Natural gas wells produce much less water than do oil wells. Depending on local geology, produced water may contain significant amounts of various mineral salts as well as organic and inorganic chemicals and radionuclides. The introduction of metals and organic compounds from produced water into water supplies is an environmental concern, occurring both through downhole escape into groundwater and by surface runoff caused by precipitation. It is also common for produced water to be high in saline concentration—in some locations it can be as high as 20 percent by weight (Stephenson, 1992). The release of this water can result in salinity levels too high to sustain plant growth and could render water supplies unusable for human consumption.

Solid waste is also produced by the settling of particles while oil is temporarily stored in on-site tanks. The tank bottoms must be cleaned periodically and the sediment, containing oil and smaller amounts of other contaminants, disposed of.

During processing, additional waste is generated from the dehydration and sweetening (further refining and conditioning) processes. Triethylene glycol is used and re-used as a desiccant. The glycols are volatile and hazardous if inhaled and thus pose a threat when they eventually are disposed of. Wastes from sweetening include spent amine solution, iron, and elemental sulfur.

Several air emissions are associated with production. Volatile organic compounds (VOCs) are released from leaking tubing, valves, tanks, or open pits. Natural gas that is not sold or used at the well is burned off, producing carbon monoxide, nitrogen oxides, and possibly sulfur dioxide. Finally, similar to drilling, the production process requires the use of fuel combustion machinery that produces typical combustion emissions.

Maintenance

Periodic maintenance "workovers" involve repairing leaks in the casing, replacing downhole equipment, well stimulation, additional casing perforations, and painting and cleaning. For the downhole work, a rig smaller than that used for drilling is brought in. The workover process requires many of the same inputs and produces similar waste and pollutants as the drilling process.

Workover-specific pollutants, many of which are toxic, may be spilled at the surface or appear in the produced water when production resumes.

To improve the flow of fluid, accumulated salts (scale) and paraffin are removed from tubing, lines, and valves using strong acids and phosphates (U.S. Environmental Protection Agency, 1992). Corrosion inhibitors are pumped through the tubing, lines, and valves to mitigate the effects of acidic components of the formation fluid. The removed scale and paraffin waste may contain trace radionuclides.

Accidents

Accidental release of oil and gas, with associated formation by-products and injected substances, can occur during spills, blowouts, and other facility disturbances. Oil spills are the most common type of accident. In 1996, 1276 onshore facilities reported crude oil spills, totaling 131,000 gallons (American Petroleum Institute, 1998). The extent of unreported spills is unknown, though the Clean Water Act requires the reporting of spills over a certain threshold. Spills most often are small and result from leaking tanks, imperfect transfers, and leaking flowlines, valves, joints, or gauges. Drilling muds have been spilled while being offloaded and production chemical spills can occur at all points of the operation. Primary spill concerns include surface contamination, runoff to streams, and seepage into groundwater, which in turn may have dire consequences for aquatic and terrestrial life.

Facilities subject to Spill Prevention Control and Countermeasure (SPCC) regulations are required to maintain containment and diversionary structures to prevent the spill from reaching vital water systems. These structures could be berms, retention ponds, absorbent material, weirs, booms, or other barriers and systems. If preexisting structures are not in place or are not adequate, and time and location allow, bulldozers are brought in to contain the spill. Remediation approaches include microbial bioremediation, composting, landspreading, or landfilling. In remote locations, in the presence of calm winds and minimal surrounding vegetation, in-situ burning is employed (Zengel, 1999; Fingas, 1999).

Well blowouts are less common, but usually more serious. They can occur at any point in the production process, but most often result during drilling and workovers. If the formation pressure exceeds that of the drilling or workover fluid, the formation fluid, injected fluid, and downhole equipment are thrust to the surface and may ignite if a spark or flame is present. The time needed to cap and control a blowout varies from a few days to a few months. Blowouts have destroyed rigs and killed workers. They cause the release of produced water and oil, and/or drilling and workover fluids, such that possible components of concern include salt, heavy metals, and oil. The produced oil and water mixture can be spread in a wide area around the rig, possibly leaching through the soil to a fresh water aquifer or running off into nearby surface waters.

Spills and blowouts also result in air emissions. Crude oil contains organic compounds that may volatilize and be emitted before the spill can be cleaned up. In-situ burning of crude oil spills and product ignition during blowouts result in particulate and carbon monoxide emissions. In rare cases, disruptions at facilities producing impure natural gas have caused the release of hydrogen sulfide.

Waste Disposal

Waste disposal is a heavily regulated aspect of extraction, with compliance intended to minimize environmental harm. Disposal options for solid and hazardous wastes, including contaminated produced water, are becoming more limited and expensive. For remote locations, such as many of those in the Rocky Mountains, waste disposal may be especially difficult and costly. About half of all produced water undergoes a substantial treatment process and is then injected for enhanced recovery. Water is also disposed of by injection into former producing formations. A very small portion is treated and used for irrigation or livestock watering. Treatment and uses of wastewater are regulated by Clean Water Act and Safe Drinking Water Act requirements.

Solid waste produced during drilling, production, and maintenance is commonly held in an on-site reserve pit and then either buried in the pit (over two-thirds of waste) or transported off-site for disposal. Off-site disposal methods include but are not limited to (a) landfarming/landspreading, in which the products from the reserve pit are broken up, thinly applied to soil, and tilled in; (b) slurry injection of solids into salt caverns; (c) recycling to use as landfill cover, roadbed construction, dike stabilization, and in the plugging of other wells; and (d) disposal in municipal or commercial landfills in areas with less oil and gas activity. Although these activities are regulated, the potential for leaching of various contaminants into the soil, surface, and groundwater remains.

Orphan Wells

When production is permanently stopped, a well is abandoned. Well abandonment is a highly regulated process that involves plugging the wellbore with cement to prevent migration of formation fluids into fresh-water aquifers.

Note that well abandonment does not mean removal of the access roads and pipelines. Wells for which production has ceased and for which no responsible party exists (because the operator either is unknown or has gone bankrupt) are known as orphan wells. Orphan wells are not properly abandoned and have the potential to contaminate local freshwater aquifers with highly saline formation water. Although not all orphan wells cause pollution, they are numerous. Approximately 134,000 wells drilled in the United States by 1995 (5 percent of all wells) are classified as orphan (Interstate Oil and Gas Compact Commission, 1996).

Transport

Pipelines and access roads broaden the range of land use issues. Equipment and waste are most often transported to and from sites via roads. Building and maintaining roads has the potential to alter or destroy habitat, degrade surface water quality and air quality, and create noise. These impacts are related to regional ecosystem problems for airsheds, watersheds, and animal migration corridors. As described above, recent studies are beginning to document various ecological impacts associated with road development in the United States. The construction of new roads can also induce environmental disturbance caused by future development unrelated to the oil and gas industry. Pipeline infrastructure also opens the land to development, potentially restricts animal movements, and exposes a broader area to the risk of oil and gas spills caused by leaks in the pipeline.

Environmental Considerations and the Viable Resource

As described above, oil and gas extraction activities are regulated to guard against environmental impacts associated with air emissions, water pollution, and solid and hazardous waste. However, regulation does not necessarily prevent projects that might result in significant environmental impacts,[5] nor is compliance guaranteed. Previous studies have incorporated lease restrictions into resource estimates of oil and gas. Our goal will be to characterize the sensitivity of lands to varying degrees of activity in order to inform the development of useful decision rules irrespective of current lease stipulations.

---

[5]Federal agencies (primarily BLM and Forest Service) are responsible for defining thresholds of significance and overseeing project approval. In granting a permit, consideration of environmental issues is tempered by consideration of other social benefits.

This type of analysis requires (1) assessing the state of relevant environment systems, (2) cataloging potential impacts of oil and gas extraction processes, (3) classifying areas according to the sensitivity of existing conditions to change, and (4) estimating gas and oil resources on lands of varying degrees of sensitivity to these impacts.

Essentially, this analysis begins by considering various sets of relevant spatially oriented environmental indicator data—appropriate "state" data and "stressor" data.[6] These data can be combined to create several spatially oriented "sensitivity indices" for various areas, with respect to various oil and gas extraction processes. Useful environmental measures coincide with those impacts associated with extraction, and would include those related to air quality, water quality (both surface and ground), soil condition and quality, physical land characteristics, and hazardous material handling. More integrative measures of biological and ecological structure and function will be included as well, to the extent data are available. Together, individual measures of air, water, and land indicate stresses exerted on ecological systems. Ecological measures indicate the change in state, or lack thereof, associated with these stresses.

Once developed, these indices will allow us to determine the degree of sensitivity of included geographical areas that coincide with underground oil and gas resources. By applying certain thresholds of significance to this assessment and considering available mitigation options at various economic costs, the effect of varying assumed levels of environmental impacts on the size of the viable oil and gas resource may be determined.

Assuming an appropriate level of resolution of available data, some additional questions may be answered by considering location of environmental sensitivities within areas of oil and gas resources. For example, if oil and gas resources occur beneath an environmentally sensitive area, is it possible to access

---

[6]"State" data indicate the status of existing systems, while "stressor" data describe the quantities of materials ("stressors") that potentially impact those systems. Many data, often in GIS or other spatial format are available, and are regularly updated as part of federal monitoring programs: (1) Data provided by the EPA exemplify useful "stressors" potentially affecting ecological resources. EPA compiles air quality monitoring data nationwide for air pollutants regulated by the Clean Air Act, compiles water quality data for watersheds around the United States as part of the Clean Water Act requirements, and maintains the Toxic Release Inventory established under the Emergency Planning and Community Right-to-Know Act of 1986 and expanded by the Pollution Prevention Act of 1990. (2) Wetlands and riparian areas tend to be useful areas to focus analyses, because their function often depends on a number of factors related to air, water, and soil, and they often support the widest range of wildlife. The Department of Interior's National Wetland Inventory and USGS mapping of streams show location of riparian and wetland habitats. (3) Occurrence of wildlife and extent of habitat are common ecological "state" indicators. The U.S. Fish and Wildlife Service and parallel state agencies maintain useful databases on occurrences of protected species, migration corridors, and ecologically sensitive areas. (4) Another useful "state" measure is that of land use, which can be tracked by satellite. The Landsat system, for example, regularly acquires landscape-scale imagery that clearly shows areas of land use change (especially resulting from development).

these resources from adjacent areas using horizontal drilling? If so, to what extent, and at what cost? Additionally, if oil and gas resources occur beneath an environmentally sensitive area, but sensitivity occurs only during particular seasons or with certain methods and mitigation options, is it possible to access these resources during the off-season or with other methods and mitigation options? Previous studies have begun to answer this last question, but timing assumptions may not have allowed certain options such as multi-year drilling, nor do they include cost of mitigation. As part of this project, we will compile data to create additional GIS layers that allow us to address these questions and further refine the viable oil and gas resource estimate.

7. Conclusions

A fundamental unanswered question in the debate over access to gas and oil resources on federal lands in the Intermountain West is how much usable resource is available at what cost and with what impact. Existing resource assessments, which estimate the technically recoverable resource, are of limited help in addressing this question because much of the resource estimated in these assessments is not likely to be produced in the next 20 years. The technically recoverable resource is not intended to account for additional factors that may impede or prevent resource production.

In an effort to provide policymakers with the information needed to understand the amount of resource that is likely to be produced, and to understand the costs and environmental impacts of such production, we have introduced the concept of the viable resource. The viable resource is a subset of the technically recoverable resource, and is that fraction that satisfies the additional criteria of being economically feasible to produce, supported by sufficient infrastructure, and for which the environmental impacts are sufficiently small or could be mitigated.

For the purposes of making policy decisions regarding gas and oil development, it is this viable resource that is of interest. Application of the viable resource criteria is meant to guide evaluation of how much resource is available at what cost and with what impact. This report outlines a number of factors that should be considered in constructing the viable resource criteria. These build upon existing work and identify important modifications that will improve their relevance.

Implications of Viable Resource Approach

The approach of defining a viable resource assessment outlined in this report has two important implications. The first is that it illustrates that the debate over federal land access needs to be refocused. Much energy has been and continues to be expended on efforts to quantify the amount of gas and oil resources in the Intermountain West that are subject to various levels of access restrictions. However, these analyses are based upon the technically recoverable resource and hence cannot address the question of how much resource is likely to be produced if access were not an issue. In addition, these analyses address federal lands only. There is public benefit in broadening the debate to address the full implications of resource development and consider the resource that is viable to produce. A debate about access restrictions alone does not fully illuminate the discussion.

The second implication is that it would be prudent to have a better understanding of the economic costs, infrastructure requirements, and environmental impacts of increased production as decisions are made with regard to changing the status of federal lands available for exploration. Increased exploration may help refine estimates of the technically recoverable resource. However, it is important to also further our understanding of the economics, infrastructure limitations, and environmental impacts surrounding increased production. These issues can be addressed through the methods proposed in this report. As understanding of the viable resource improves, decisions about increasing exploration and production can be better focused through an enhanced understanding of the benefits and impacts of those efforts.

Potential Results

The criteria presented in this report are meant to provide the basis for further work to quantitatively evaluate the viable gas and oil resource in the Intermountain West, with particular attention on the Greater Green River Basin. Nonetheless, based on existing work, we can begin to understand the effect of the viability criteria on the amount of gas available in the Rocky Mountain Region.

Figure 7.1, based on the results of the U.S. Geological Survey economic analysis (Attanasi, 1998), shows the effect of the wellhead economic criterion in the Rocky Mountain Region and Greater Green River Basin. This figure shows that the economic criterion alone can have a substantial impact on the size of the viable resource. According to the U.S. Geological Survey, at a wellhead price of $3.34 per thousand cubic feet of gas (equivalent to $30 per barrel of oil), less than 20 percent of the technically recoverable gas in the total Rocky Mountain Region is economically recoverable, and only 5 percent of the technically recoverable gas in the Greater Green River Basin is economically recoverable. Note that these results do not reflect RAND's analysis. The costs of exploring and developing gas and oil deposits in the Rocky Mountain Region are decreasing with technological advances. Our economic analysis will use different data and assumptions and may produce different results.

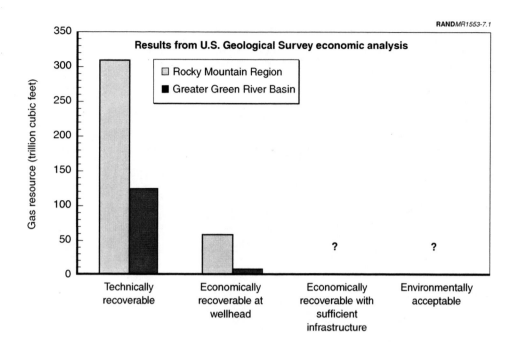

Figure 7.1—Potential Effect of Viability Criteria on Gas Resources

Future Work

This report lays the foundation for a methodology to define the viable gas and oil resource. The next step will be to refine this methodology and apply it to individual basins in the Intermountain West, starting with the Green River Basin. This will entail developing relationships among gas and oil deposit characteristics, technology options, infrastructure requirements, environmental impacts, various costs, and other variables to allow us to quantitatively assess the amount of resource that satisfies specified viability criteria. It will also entail gathering the relevant data for each area being considered.

We envision a model in which the user is able to choose values for key variables, such as resource prices, cost components, drilling technologies, and levels of different environmental impacts. Outputs will be presented both numerically as well as in the form of Geographic Information System layers showing the amount and location of resources that satisfy the various viability criteria. Such an output will provide a simple way to characterize the viable resource in the context of the deposit types, well locations, existing and necessary new infrastructure, environmental sensitivities, topography, and any other of a number of relevant spatial attributes in the area of interest. This way of presenting resource assessments would be a substantial enhancement of present information.

Appendix

Technically Recoverable Resource Assessment Specifications

The specifications of the different resource assessments used for this analysis vary. It is important to understand how these specifications differ to correctly compare the assessments. Table A.1 provides a summary of the specifications by assessment.

Table A.1

Comparison of Resource Assessment Specifications

	NPC	USGS/MMS	PGC
Date Released	December 1999	USGS: 1995 MMS: 2000	March 2001
End date of assessment	January 1, 1998	USGS: January 1, 1994 MMS: January 1, 1999	December 31, 2000
Commodities	Natural gas	Crude oil, natural gas, natural gas liquids	Natural gas
Resource categories	Proved reserves Reserve appreciation Undiscovered conventional Nonconventional	USGS: Reserve appreciation Undiscovered conventional Nonconventional MMS: Proved reserves Unproved reserves Reserve appreciation Undiscovered conventional	Reserve appreciation Undiscovered conventional Nonconventional (coalbed methane only)
Major regions assessed	Lower 48, onshore and offshore (state and federal waters), Alaska, Canada	USGS: Onshore and state waters for lower 48 and Alaska MMS: Federal offshore waters in lower 48 and Alaska	Lower 48, onshore and offshore (state and federal waters), Alaska

Table A.1 (continued)

	NPC	USGS/MMS	PGC
Subregions	Lower 48: 13 onshore and 4 offshore regions; onshore regions divided into 4 depth intervals (0–5, 5–10, 10–15, >15 thousand feet); offshore regions divided into 8 water depths or areas that vary by basin Canada: 5 regions, each region characterized by drilling depth and/or subregional area	USGS: 8 regions consisting of 71 provinces and 560 individual plays; 100 are nonconventional MMS: 4 offshore regions; 2 geological provinces (Mesozoic and Cenozoic); 103 plays, including conceptual and established	7 areas or regions consisting of 89 geological provinces; onshore provinces are divided by depth (<15 and 15–30 thousand feet); offshore divided by water depth (<200 and 200-1000 meters)
Format	Most likely estimate	Risking structure with probability estimates ranging from 5 percent to 95 percent	Probable, possible, speculative, and most likely estimates

Date Released

The National Petroleum Council (NPC) assessment was completed as a one-time follow-up to a 1992 NPC natural gas study. It was released in December 1999 with the work initiated at the request of the Secretary of Energy in May 1998. The Potential Gas Committee (PGC) assessment is completed on a biennial basis. The 1999-2000 report was released in March 2001. The work was completed during late 1999 and in 2000. The U.S. Geological Survey (USGS) assessments have been completed systematically since 1975. The USGS assessment, released in 1995, was completed between 1991 and 1994 as part of a joint USGS/Minerals Management Service (MMS) assessment of the U.S. oil, natural gas, and natural gas liquid resource. The MMS released a revised assessment in 2000. The work for this revised assessment was completed during 1999. The revised MMS assessment was used for this analysis.

End Date of Assessment

A portion of proved reserves are produced and delivered for consumption each year. As a result, the cumulative production figure increases each year. Typically, a percentage of the potential resource (reserve appreciation, new field undiscovered, and nonconventional) are confirmed and added to proved reserves. As a result, proved reserves do not decline by an amount equal to production each year. As such, resource assessments are dynamic. An assessment with an end date of January 1, 1994 (USGS), cannot be directly compared to an assessment with an end date of December 31, 2000 (PGC). The resource assessments need to be adjusted for production and net-proved-reserve additions to some common date. For example, between January 1, 1994 (the end date of the USGS assessment), and December 31, 2000 (the end date of the PGC assessment), 110 tcf of natural gas was produced or added to proved reserves in the lower-48 onshore. This total needs to be apportioned and subtracted from the USGS potential resource estimate to provide a valid comparison with the PGC assessment. Not making this correction is a common mistake made when comparing two resource assessments. For purposes of this analysis, all of the resource assessments presented were adjusted to a common end date of December 31, 2000.

Commodities

The NPC and PGC assessments only covered natural gas. The Gas Research Institute (GRI) Hydrocarbon Supply Model, which was used for the NPC study, considers natural gas, crude oil, and natural gas liquids. However, only the natural gas resource assessment was presented in the NPC documentation. The USGS and MMS assessments cover a broader group of commodities, including natural gas, crude oil, and natural gas liquids.

Resource Categories

All of the assessments cover a similar group of resource categories. The terminology varies, but the definitions are very similar. Table A.2 provides a comparison using the terminology used in each resource assessment.

The key difference between the resource categories is coverage. The NPC and USGS assessments provide a proved reserve figure, but it is simply the EIA figure and not an original assessment. The PGC assessment does not include an estimate of proved reserves, instead focusing on the unconfirmed or potential resource categories. The MMS is the only assessment that provides an original proved reserve estimate. However, the difference between the EIA reported proved reserve total and the MMS figure is small. For purposes of this analysis, the EIA proved reserve figure was used to improve comparability. With the exception of differences in terminology, all of the assessments include an estimate of the reserve appreciation, new field undiscovered, and nonconventional resource. There are, however, some key differences in

Table A.2

Comparison of Resource Categories

	NPC	USGS	MMS	PGC
Proved reserves	Reported EIA[a] data	Reported EIA data	Proved reserves (original estimate)	Not assessed
Reserve appreciation	Old field reserve appreciation	Reserve appreciation in conventional fields	Reserve appreciation + unproved reserves	Probable resource
New field undiscovered	New fields	Undiscovered conventional resources	Undiscovered conventionally recoverable resources	Possible + speculative resources
Nonconventional	Tight, shale, coalbed methane	Continuous-type accumulations in sandstone, shale, and chalks and continuous-type accumulations in coal beds	Not assessed	Coalbed methane
Cumulative production	Cumulative production (estimated)	Not assessed	Not assessed	Not assessed

[a]Energy Information Administration.

coverage of nonconventional resources. The NPC assessment includes separate estimates for coalbed methane, shale gas, and tight gas. The USGS assessment includes two categories: continuous-type accumulations in sandstone (tight gas), shale, and chalk; and continuous-type accumulations in coal beds. The PGC considers only coalbed methane nonconventional resources. The MMS includes no assessment of nonconventional resources in the offshore.

Major Regions Assessed

The NPC assessment includes both the onshore and offshore (state and federal waters) in the lower-48, Alaska, and Canada. The PGC assessment covers only the lower-48 and Alaska. The USGS resource assessment covers the onshore and state waters, but excludes resources in the federal offshore waters. The MMS assessment covers only the federal offshore waters (outer continental shelf). For purposes of this analysis, the USGS and MMS assessments were combined to provide a complete U.S. assessment for the lower 48 states.

Subregions

The regional detail level provided in each of the assessments varies considerably. In general, the NPC assessment provides information at the regional level, PGC at the province or basin level, and USGS and MMS at the play level. While the regional detail varies, it is sufficient to allow comparison on a consistent regional basis for the Rocky Mountain Region as will be discussed later.

The NPC assessment includes 13 onshore and 4 offshore regions. The onshore regions are divided into 4 depth intervals: 0 to 5000 feet, 5000 to 10,000 feet, 10,000 to 15,000 feet, and greater than 15,000 feet. The offshore regions are divided into up to eight water depths or areas. The applicable depth intervals or areas vary by offshore basin. The PGC assessment includes seven areas or regions that consist of 89 geological provinces or basins. The onshore provinces are divided into depth intervals, less than 15,000 and 15,000 to 30,000 feet. The offshore is divided by water depth less than 200 meters and 200 to 1000 meters. The USGS assessment includes eight regions, which consist of 71 provinces. The USGS assessment is unique in presenting the play-level detail: 560 plays were assessed, including 100 nonconventional or continuous deposits. The detail level in the MMS assessment is similar to that in the USGS assessment. The MMS covers four offshore regions. It includes two geological provinces (Mesozoic and Cenozoic) and 103 plays (including conceptual and established plays) that are assessed within the regions and provinces.

Format

Many people assume that a resource assessment is the equivalent of a complete inventory of the particular resource being assessed. However, this is not the case. As noted previously, the quantities being evaluated are largely unknown and it is impossible to measure them precisely. The assessments are an attempt to bound that uncertainty. All of the assessments employ some type of probabilistic format consistent with the inherent uncertainty of the assessments.

The NPC resource assessment provides a single resource number. However, that assessment is defined as a most likely estimate, implying the potential that the total could be smaller or larger. The range of that uncertainty is not explored in the NPC documentation. The PGC assessment defines a minimum, most likely, and maximum estimate for each resource category (e.g., probable, possible, and

72

speculative). While these are not probabilities, each estimate is associated with a probability. The minimum estimate assumes the existence of a minimum number of traps, that many of the traps will not contain recoverable gas accumulations, and a minimum yield factor for each basin. The minimum estimate is associated with close to a 100 percent probability that at least this much gas resource is present. The most likely estimate is consistent with the estimator's best judgment of the number of traps, yield factor, and reservoir conditions. This is the equivalent of the most likely estimate (mean estimate) of the resource. The maximum estimate is associated with the most favorable conditions (maximum number of traps, yield factor, and reservoir conditions). The maximum estimate has close to a zero probability of occurrence. The USGS and MMS resource assessments define risk structures in a similar fashion to that used by PGC. Their risk structures are based on three attributes: charge, reservoir, and trap and are applied at the play level. The USGS and MMS resource assessments are provided at 95 percent, mean, and 5 percent probability levels.

References

Advanced Resources International, Inc., *Undiscovered Natural Gas and Petroleum Resources Beneath Inventoried Roadless and Special Designated Areas on Forest Service Lands Analysis and Results*, 2000.

Advanced Resources International, Inc., *Federal Lands Analysis Natural Gas Assessment Southern Wyoming and Northwestern Colorado Study Methodology and Results*, http://fossil.energy.gov/oil_gas/reports/fla/, 2001.

American Gas Association, *How Does the Natural Gas Distribution System Work?* www.aga.org/Newsroom/InfrastructureSecurity/4235.html, 2001.

American Petroleum Institute, *Petroleum Industry Environmental Performance Sixth Annual Report*, American Petroleum Institute, 1998.

American Petroleum Institute, *Overview Of Exploration And Production Waste Volumes and Waste Management Practices in the United States*, American Petroleum Institute, 2000.

Attanasi, Emil D., *Economics and the 1995 Assessment of United States Oil and Gas Resources*, U.S. Geological Survey Circular 1145, 1998.

Barlow and Haun, Inc., *Accessibility to the Greater Green River Basin—Gas Supply, Southwestern Wyoming*, Gas Research Institute, Report GRI-94/0363, 1994.

Berger, Bill D., and Kenneth E. Anderson, *Modern Petroleum—A Basic Primer of the Industry, 3rd edition*, PennWell Books, Tulsa, OK, 1992.

Bureau of Land Management, Pinedale Field Office, Faxed data. www.wy.blm.gov/pfo/wildlife.htm, 2002.

Defenders of Wildlife, *Seriously Flawed Energy Department Report Threatens Reasonable Wildlife Protections on Public Lands–More Reports Planned*, www.defenders.org/publiclands/doereport.pdf, 2001.

Demographia, Area, Population and Density by U.S. State, 1990, www.demographia.com/db-landstatepop.htm, 2002.

Department of Energy Office of Fossil Energy, *Environmental Benefits of Advanced Oil and Gas Exploration and Production Technology*, DOE-FE-0385, 1999.

DuVall, S. L., *Federal Land Access to Oil and Gas Minerals in Eight Western States*, Cooperating Associations Forum, 1997.

East of Huajatolla Citizens Alliance, *Well Spacing*, www.ehcitizens.org/cbmgas/is11_spacing.htm, 2002.

74

Energy Information Administration, "Status of Natural Gas Pipeline System Capacity Entering the 2000-2001 Heating Season," in *Natural Gas Monthly*, October 2000, DOE/EIA-0130(2000/10), 2000.

Energy Information Administration, *Historical Natural Gas Annual 1930 Through 2000*, DOE/EIA-E-0110(00), 2001.

Energy Information Administration, "The Rocky Mountains–A Persian Gulf for Natural Gas?" in *Performance Profiles of Major Energy Producers 2000*, DOE/EIA-0206(00), 2002.

Ercelawn, A., *End of the Road: The Adverse Ecological Impacts of Roads and Logging: A Compilation of Independently Reviewed Research*, Natural Resource Defense Council, New York, 1999.

Fingas, M. L., *In Situ Burning of Oil Spills: A Historical Perspective*, in *In Situ Burning of Oil Spills Workshop Proceedings, New Orleans, Louisiana, November 2-4, 1998*, William D. Walton and Nora H. Jason, eds., Building and Fire Research Laboratory, National Institute of Standards and Technology, Gaithersburg, MD, 1999.

Holtberg, P. D., *Gas Demand & Supply Overview: Getting Past the Near Term Difficulties*, Presentation to Senior Scientists & Engineers, Washington Chapter, www.seniorscientist.org/SSEPres1/index.htm, March 2000.

Houghton, J. T., G. J. Jenkins, and J.J. Ephraums (eds.), *Climate Change: The IPCC Scientific Assessment*, Cambridge University Press, 1990.

LaTourrette, T., M. Bernstein, P. Holtberg, C. Pernin, B. Vollaard, M. Hanson, K. Anderson, and D. Knopman, *A New Approach to Assessing Gas and Oil Resources in the Intermountain West*, IP-225-WFHF, RAND, Santa Monica, CA, 2002.

Mac, M. J., P. A. Opler, C. E. Puckett Haecker, and P. D. Doran, *Status and Trends of Our Nation's Biological Resources*, U.S. Geological Survey, Reston, VA, 1998.

Minerals Management Service, *Outer Continental Shelf Petroleum Assessment, 2000*, U.S. Minerals Management Service, 2000.

Morton, Peter, *The Department of Energy's "Federal Lands Analysis Natural Gas Assessment," A Case of Expediency over Science*, www.wilderness.org/newsroom/pdf/doe_greenriver_071001.pdf, The Wilderness Society, 2001.

National Petroleum Council, *Potential for Natural Gas in the United States*, National Petroleum Council, 1992.

National Petroleum Council, *Natural Gas: Meeting the Challenges of the Nation's Growing Natural Gas Demand*, National Petroleum Council, 1999.

Potential Gas Committee, *Potential Supply of Natural Gas in the United States*, Potential Gas Agency, Golden, CO, 2001.

75

Root, David H., Emil Attanasi, Richard F. Mast, and Donald L. Gautier, *Estimates of Inferred Reserves for the 1995 USGS National Oil and Gas Resource Assessment*, U.S. Geological Survey Open-File Report 95-75L, 1997.

Sittig, Marshall, *Petroleum Transportation and Production: Oil Spill and Pollution Control*, Noyes Data Corporation, Park Ridge, NJ, 1978.

Stephenson, M. T., "A Survey of Produced Water Studies," in *Produced Water: Technological/Environmental Issues and Solutions*, International Produced Water Symposium, James P. Ray and F. Rainer Engelhardt, eds., Plenum Press, New York, 1992.

U.S. Environmental Protection Agency, *Management of Wastes from Oil and Gas Exploration, Development, and Production*, Report to Congress, Office of Solid Waste, U.S. Environmental Protection Agency, Washington, D.C., 1987.

U.S. Environmental Protection Agency, *Background for NEPA Reviewers: Crude Oil and Natural Gas Exploration, Development, and Production*, Office of Solid Waste, U.S. Environmental Protection Agency, Washington, D.C., 1992.

U.S. Environmental Protection Agency, *EPA Office of Compliance Sector Notebook Project: Profile of the Oil and Gas Extraction Industry*, EPA/310-R-99-006, U.S. Environmental Protection Agency, Washington, D.C., 2000.

U.S. Geological Survey National Oil and Gas Resource Assessment Team, *1995 National Assessment of United States Oil and Gas Resources*, U.S. Geological Survey Circular 1118, 1995.

U.S. Geological Survey, *Global Land Information System*, USGS Digital Elevation Model Data, edcwww.cr.usgs.gov/glis/hyper/guide/usgs_dem, 2002.

Vidas, E. Harry, Robert H. Hugman, and David S. Haverkamp, *Guide to the Hydrocarbon Supply Model: 1993 Update*, Gas Research Institute, Report GRI-93/0454, 1993.

Zengel, Scott A., et al., "Environmental Effects of In Situ Burning of Oil Spills in Inland and Upland Habitats," in *In Situ Burning of Oil Spills Workshop Proceedings, New Orleans, Louisiana, November 2–4, 1998*, William D. Walton and Nora H. Jason, eds., Building and Fire Research Laboratory, National Institute of Standards and Technology, Gaithersburg, MD, 1999.